THE FUN OF RAISING A COLT

Frank B. Griffith
and
Rubye Mae Griffith

1979 Edition

Published by
Melvin Powers
WILSHIRE BOOK COMPANY
12015 Sherman Road
No. Hollywood, California 91605
Telephone: (213) 875-1711

© 1970 by A. S. Barnes & Co., Inc.
Library of Congress Catalogue Card Number: 72-88266

Printed by
HAL LEIGHTON PRINTING COMPANY
P.O. Box 3952
North Hollywood, California 91605
Telephone: (213) 983-1105

Printed in the United States of America

ISBN 0-87980-190-5

CONTENTS

1	Why Raise a Colt?	9
2	To Market, To Market, to Buy a Fine Mare	16
3	Choosing a Suitable Sire	45
4	The Long Wait—Watching and Wondering	55
5	As the Big Day Approaches	73
6	New Arrival	93
7	Bringing Up Baby	118
8	Early Discipline	134
9	More Advanced Training	149
10	At the Show	162

1

WHY RAISE A COLT?

SOME EXCELLENT REASONS FOR RAISING A COLT

Every horse lover will tell you that horses are fun. But the most fun of all is a colt! Especially one you raise yourself.

And besides all the fun that goes with raising a colt there are so many additional rewards. For one thing, it's really the best way to learn to understand horses. When you're around a horse from the time it's foaled you can observe all its follies and foibles, and so you get to know what makes horses tick. Also, raising a colt is a wonderful way to gain experience in the training and handling of horses. And of course there's something about the relationship that develops between young owner and young horse that somehow refines the character of both. Then too, and we consider this one of the most exciting rewards of colt raising, you get to meet so many nice people—and horses!

Reviewing all these good reasons for raising a colt we'll settle for the first one: pure fun.

POSITIVE SIDE OF COLT (FOSTER) PARENTHOOD

Selecting the sire (father) and dam (mother); watch-

Two young horse lovers experiencing the unadulterated joys of horse ownership. (Courtesy Vivian Cook and Sandy Holmes, Agoura, California.)

WHY RAISE A COLT?

ing and waiting for the youngster to arrive; witnessing the miracle of birth; tending the young foal as it takes its first tentative steps toward independence; and applying the principles of training and discipline: all these things add up to one of the richest experiences in a horse lover's life. But they represent only part of the picture. There's another side to the coin and we want to consider it right now.

AND CERTAIN OTHER CONSIDERATIONS OF COLT RAISING

Besides the pure pleasure of raising a colt you must keep in mind the extensive time investment involved.

Counting the gestation period of the mare, which is usually a minimum of 336 days, and the length of time required to bring the colt to a usable or marketable age, you'll find you may have invested as much as three or four years in your venture. This is true because the foal as a yearling is still a playful creature, considered unfinished from a marketing viewpoint.

At two he should not be ridden too seriously or be subjected to any really heavy weight, particularly if he is a slightly built colt. So it isn't until he is at least three that he may even be considered "green broke." By "green broke" we mean able to be ridden with a youth bit but not broken to neck rein, to back up, or to perform any of the other niceties of finished training.

SUSTAINED INTEREST AN IMPORTANT FACTOR

With a time investment this prolonged, it is very important for you to feel that you can sustain interest in your colt-raising project over an extended period of time.

Those who oh and ah at the thought of having a darling little gangling colt to fondle and who look forward with eagerness to raising the little rascal must be warned that they face months and months of hard work, as well as fun; and it won't be possible to walk away from the project when the going gets tough—which it will.

MORE REALISTIC DETAILS

Another thing you have to be sure of when you decide to raise a colt of your own is the need for at least a rudimentary knowledge of horse care. Either you must have this knowledge, a member of your family must be able to supply it, or you must have access to friendly, competent advisors who will be willing and able to see you through periods of crises.

SUITABLE QUARTERS FOR MOMMA

Both mother and foal require proper quarters, and you must be able to provide them. The mare should have a clean, warm, dry, spacious, and sheltered area or stall in which to have her foal. And she must be kept up (that is, indoors or protected from inclement weather) while she is expecting.

She will also need adequate exercise space.

And she will have to be properly fed and watered and groomed.

You will discover that you also need all sorts of special equipment for both mother and foal during the long period of colt raising.

All of these requirements will be gone into in greater detail below. We mention them briefly now so that the fainthearted may drop out while there is still time.

MONEY MATTERS

One other possibly painful point we should cover is the fact that this investment of time and equipment involves a considerable expenditure of cold cash. Some of

The owner (astride), her good friends, and even the dog all share the pleasure of a trustworthy pleasure horse. (Courtesy Sandy Holmes, Agoura, California.)

Four hearts that beat as one. Is there any doubt of the two-way love felt here? (Courtesy Martin Moran and Stephen and Warren Greer.)

your expenses may be financed but this will have to be worked out with your family or your horse club leaders.

When you get down to final details you should estimate your investment in time and money realistically. Your costs will include purchase of the brood mare, inspection

WHY RAISE A COLT?

and breeding fees (including transportation to the stallion), and boarding fees for the mare during the thirty day period that you must wait to see if she is "settled," that is, pregnant.

You'll also have the expense of quarters and equipment just mentioned, some vet and doctoring bills—almost certainly—and of course feed bills for mother and colt over the one, two, or three year period during which the colt is reaching maturity.

Does it sound rather awesome and staggering?

ACCENTUATE THE POSITIVE

If you really want to raise that colt badly enough and you can fix your eye on the fun and not on the problems, you'll find the ways and the means to carry you through your project and the rewards, as we said, though not monetary, will more than repay you.

So take a deep breath, hold to your goal, picture that frisky, frolicsome creature, with tossing mane and flag-high tail, galloping toward you, sliding to a halt at your side, and nuzzling your hand with its velvet-soft muzzle, and nothing on earth will stop you from enjoying the indescribable joy and pure pleasure of raising your very own colt!

2

TO MARKET, TO MARKET, TO BUY A FINE MARE

PEDIGREED PARENTS OR JUST PLAIN FOLKS?

Until you have your own brood mare and raise your own colt you really haven't tasted the fullest joys of horse ownership. We heartily recommend that you indulge yourself in this greatest of luxuries.

And when we say luxury we don't mean that you must necessarily spend a vast amount of money on your brood mare. Nor must you become involved in exorbitant breeding fees when you look around for a papa.

Of course if you plan to sell your colt, you will get more for it if it comes of registered stock. But you can have just as much fun raising a colt from a grade (non-registered) mare—or for that matter a colt from a donkey or burro—as you can raising a colt out of costly thoroughbred parents.

Humans can't choose their relatives. But you can choose the parents of your colt. And to have a really fine colt you must start with the finest possible sire and dam, though not necessarily the most expensive.

A DISCUSSION OF BREEDS

The first step in the selection of parents is to look

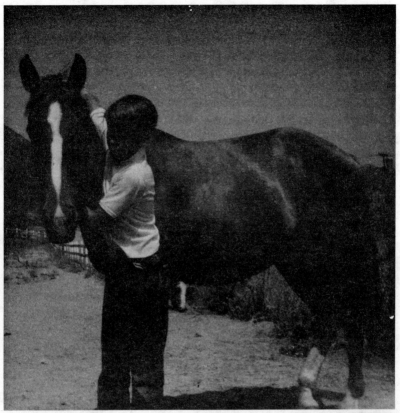

A boy and his colt. Surely all the work, all the time, all the effort of colt raising pay off in moments like these. (Courtesy Eddie Burza.)

around for a suitable brood mare. And before you decide on a particular mare you should give quite a bit of thought to the specific breed that appeals to you.

Three or four breeds continue to bring top prices over most of the country.

These are the Quarter Horses, the Arabians, the Appaloosas and the Morgans. But the popularity of breeds varies from time to time and in different localities, so it isn't always possible to predict which breed will be on the way up and which on the way down as far as market value goes.

Since you will be making a considerable investment of time and effort in your colt raising project, it won't do to skimp on the selection of parents. If the market value of the colt is important, you should inquire among local horse dealers, horse owners, horse breeders, breed associations, and horse club leaders to find out exactly which breeds are bringing the best price on the open market.

SOME POPULAR BREEDS TO CONSIDER

The following breeds of pleasure horses have remained popular for a number of years in many parts of the United States and so may be counted on to bring higher than average prices if the "get," that is, the offspring, turn out to be top quality.

QUARTER HORSES

This breed ranks high in market value almost everywhere because of handsome appearance, excellent performance records, and trustworthy disposition. Registered stock, if desirable, bring anywhere from $1000 to $50,000, depending on ancestry, conformation, and action in the show ring or on the range.

AMERICAN SADDLE BRED HORSES

This is a breed with flash and class that maintains its popularity in both the East and the South. Prices may be equal to those obtained for Quarter Horses depending on blood lines, etc.

PALOMINOS

This is a color line, which means that the color may breed true if you hit the right gene combination. With color lines it is possible to breed a registered sire to a nonregistered, or grade, mare and produce a foal that can be registered. Queries to the breed associations will explain how this is done. Prices vary, depending on quality of stock and location and are usually higher in the west and south.

APPALOOSAS

Again, this is a color line, breeding true to two favored patterns: the allover spotting referred to as *leopards,* and the rump, or withers-to-rump markings referred to as the *blanket types.* As with Palominos, you can breed a registered sire to a nonregistered mare and register the foal if the color is loud enough to be seen from the judges' stand. Also the colt must show all the other breed characteristics including the sclera around the eye, striated hoofs, mottled skin, and scanty mane and tail.

Many other breeds including Pintos, Morgans, Morabs (half Arabian and half Morgan) also provide excellent parent stock for colt raising projects. But at least one parent should be registered if you hope to command a worthwhile price for your foal.

STUD FEES

Stud fees vary across the country and are determined by quality of stock and location. The lowest price for any reputable registered stud is seldom less than $100. And stud fees for Arabians and Thoroughbreds may range from $500 to $3500 and up.

We have delayed reference to the Arabian and Thor-

Appaloosas: (a) solid bay mare bred to (b) loud color Appaloosa stallion produces (c) loud color foal that may be registered as an Appaloosa.

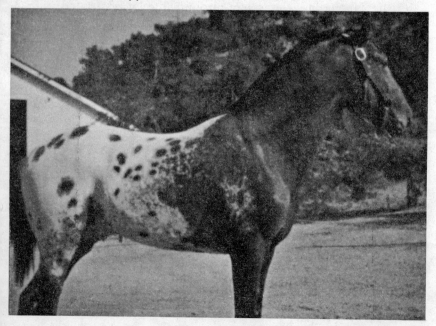

oughbred breeds because while they are both highly popular breeds and produce magnificent offspring that sell for top prices, they are generally too high-priced for the average junior colt-raising budget.

However, if you would love very much to raise a foal with at least some percentage of Thoroughbred ancestry, and you must still count pennies, we have a suggestion for you.

BREEDING BARGAINS

Go to a reputable Thoroughbred Horse Auction making sure that you take a horse expert with you. This latter advice is vital since you should never go to any horse auction with the hope of securing a bargain unless you really know your horseflesh. In fact, we caution you never to buy any horse at auction without an expert's advice. But if you do take an expert with you, it may be possible to pick up an older or even slightly unsound mare and use her for breeding purposes.

It isn't always possible to secure registration papers for a mare purchased this way since owners of registered mares are reluctant to surrender their breeding certificates when the mare brings only a minimum price. However, sometimes a kindly owner, interested in promoting an interest in breeding among young horse lovers, will break this rule. It will depend on the temperament and philanthropic disposition of the mare's owner and your personal appeal as to whether or not you get papers with a bargain mare or not.

In any event, whether you obtain registration papers or forfeit them, you will have the personal satisfaction of knowing that your mare will have excellent blood lines and the beauty and excellence of her get will tell any trained horseman that your colt is a good one with fine blood lines somewhere in its background.

TO MARKET, TO MARKET . . .

Rhap-Saudi Radi at six months already displays his proud Arabian ancestry. (Courtesy Micki Robison and Pat Freimuth, Somas, California.)

MUST A BROOD MARE BE COMPLETELY SOUND?

We should add a word of explanation here regarding the use of the word *sound* in connection with a brood mare. No prospective young breeder of colts would select a mare for breeding purposes that was unsound in conformation or in poor health.

But there are types of unsoundness that will not affect

the quality of the mare's offspring but may considerably lower her purchase price. These include lameness that would not be hereditary—that is, lameness resulting from injury or accident. Also she might be getting along in years. Or she might be temporarily run down from poor feeding or overwork.

Any of these latter factors would not influence the quality of her offspring. Of course some breeders will tell you older mares are not apt to get in foal or may have difficulty foaling. This may be true in some instances but we have some facts to present on the opposite stand.

ADVANTAGES OF AN OLDER MARE

We have bred far too many older mares and had far too many successful results, without complications, to accept as dogma the belief that older mares do not breed successfully. We say, take your chances on getting an older mare in foal if her breeding appeals to you instead of taking a chance on a younger mare of lesser breeding simply because the younger mare may seem a likelier prospect for breeding. We've had young mares that did not get in foal. And we've seen mares 24 years old produce excellent offspring without any difficulty.

Sometimes a Thoroughbred mare may have sprung a tendon and been retired from the track because her racing days were ended but you can pick her up at auction at a bargain and have yourself a beautiful foal—if you're lucky. So why not try.

PONIES AS POSSIBLE PARENTS

Another thought to bear in mind in shopping for mother material is this: Welsh Ponies and Shetlands enjoy recurrent if fluctuating popularity and some of the Ponies of the Americas—these are Appaloosa ponies—command

Brother and sister share the work, share the fun of a colt-raising venture. (Courtesy Eddie and Jessica Burza, Agoura, California.)

Mare helps free minutes-old foal from amniotic sac.

very high prices because they are being raced in harness in many parts of the country.

The Ponies of the Americas bear investigating and you should contact their breed association, a very active group, to learn more about them. One parent must be a registered Appaloosa, the other parent a registered Welsh or Shetland Pony for the pony to be eligible for registration in the association. They are beautiful, graceful, colorful little beasts, and young people adore them.

So don't rule out the possibility of raising a pony colt.

Or the colt of a male mule and a horse—the product of which is a jack; or the colt of a female mule and a horse, the product of which is a jenny. If you think the average horse colt is adorable, lovable, and amusing you will discover that all these qualities are heightened in the tiny offspring of pony mothers or the whimsical get of mules, donkeys, and burros.

These little critters take up less room than a full size horse, they eat less and are often more manageable by young owners. Therefore it would be well to think long and hard before you decide whether or not you want to play foster parent to a darling little pony foal. We've raised many ourselves and can guarantee that they are an endless delight. The only thing to remember is that they can slip through almost any fence and squeeze through any barrier, so you really must take extra precautions for containing ponies if you decide to raise any.

PEDIGREES NOT ALWAYS NECESSARY

Before concluding this subject of parentage, we'd like to repeat a thought we brought up earlier, which is the fact that any mare, whether she is papered or not papered, will produce a colt you can enjoy, love, and learn from. Just as any dog, whether it happens to be a mutt or a show dog will produce a pup that will find its way into your heart. So don't be too concerned about finding the perfect papa or the most admirable mother.

If you find a mare guaranteed in foal but no one is willing or able to say how she got that way, if you strike a good bargain and you really like the mare, take your chances on the foal. It may be a wonderful way of at least getting started on your first colt-raising project without busting your budget altogether.

Fancy breeding and impressive papering can come

later. We've raised many a little "orphink" or volunteer colt that turned out to be the darling of the ranch.

CONSIDER THE HEAVIER BREEDS

While we have concentrated in this discussion on the light breeds of pleasure horses, it can be just as much fun to raise a work colt, and certainly a lot more practical if you live on a ranch or a farm. Just as all brides are beautiful and all children adorable, the offspring of the heavier breeds of work horses appear quite enchanting when they are infants. And a really well put together and well trained work colt can bring a fine price in a farm market since these breeds are growing scarcer all the while.

IMPORTANCE OF VISITING BREEDING FARMS

Once you have decided what breed of horse you would like, the real search for a mare begins.

This is the time for you to visit several breeding farms before making your final selection. And what a pleasant family excursion this can be!

Look in your local directory for the location of breed associations near you or contact your local equestrian or 4-H Club for this information. Then line up at least four or five farms within visiting range of your home. Plan an itinerary starting with the farm farthest away and visit a farm a week or a farm every two or three weeks, depending on the amount of time and the amount of family cooperation at your disposal.

The least that can happen on a happy adventure of this sort is that you'll see some beautiful countryside, meet some beautiful horses, and make some lasting friendships among dedicated horse lovers. You'll also pick up all sorts of helpful information that will prove extremely useful when you buckle down to the fun of raising a colt.

TO MARKET, TO MARKET ...

The foal in the previous photograph gets a hug from foster mother, Margie Moss.

Take along a picnic lunch and, just to show that you are being businesslike about your horse venture, include a camera so that you can take pictures of mares or stallions that appeal to you. Then you can compare these photos with others garnered along your way. You should have a handsome collection by the time you complete your exploratory trips.

The photos and related information collected should be assembled in a book that will form part of your colt-raising portfolio.

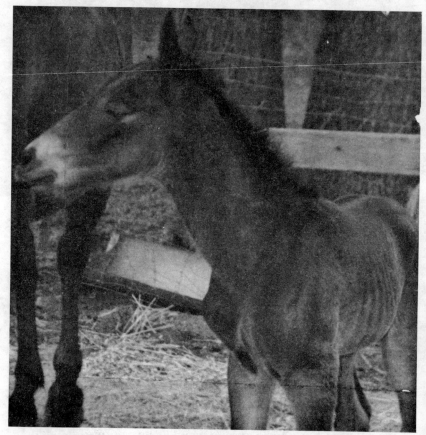

Three hours old, the new foal wonders what the world is all about . . .

DOING YOUR PAPER WORK

You should plan from the beginning to take notes on all phases of your colt project, and these notes might be kept in an attractively bound looseleaf notebook.

Your notebook should contain, among other things, several pages with the following headings, and any additional variations you may wish to add to suit your taste and particular needs.

> YOUR NAME
> NAME OF YOUR RANCH OR RESIDENCE
> DATE
> NAME AND LOCATION OF BREEDING FARM VISITED
> NAME AND TITLE OF PERSON INTERVIEWED AT BREEDING FARM
> NAMES AND REGISTRATION NUMBERS (IF ANY) OF DAMS OR SIRES
> YOUR COMMENTS REGARDING LIKES AND DISLIKES OF ANIMALS VIEWED
> SOME NOTATIONS ON THE PARTICULAR BREED SUPPLIED BY BREEDER
> NOTES ON BREEDING FEES, OPEN BREEDING DATES, BOARDING FEES, TRANSPORTATION FEES, ETC.
> ANY SPECIAL BREEDING REGULATIONS
> COMMENTS ON MANNER IN WHICH FARM IS RUN

BILL OF HEALTH FOR MARE AND STALLION

Stallion owners will want your brood mare, if brought to their stud farm, to be guaranteed "open," that is, ready for breeding and to have a clean bill of health as verified by a dated and signed veterinarian's inspection.

And if you take your mare to a breeding farm, you will want to know that the stallion is guaranteed in good health, that sanitary measures are observed during and after breeding, and that the entire establishment is run in a humane and hygienically approved manner.

. . . and Mother tells him!

SOME ADVANCE PREPARATIONS

While you are conducting your search for ideal parents for your colt, there are many things you can be doing toward the proper housing of your mare when you finally bring her home.

These will include plans for a shelter from heat, cold, and dampness, a birth place for the babe-to-be, and an exercise area for both mother and foal. You will also

TO MARKET, TO MARKET ... 33

want a setup for feeding and watering and an assortment of gear and grooming and health aids. Let's take things up in order.

THE MATTER OF SHELTER

The junior horse owner is not expected to provide Hilton-type quarters for his brood mare. The simplest lean-to, if erected with consideration for the mare's needs and if planned with forethought, will be quite adequate.

The orphaned foal drinks from a calf-feeding bottle.

The main factors to keep in mind are sturdiness and safety, each of equal importance.

Used lumber is perfectly acceptable but it should be smooth and not splintery; nor should it be coated or painted with any paint or protective substance that might be injurious to mother or foal if either should decide to nibble. And horses, both young and old, are notorious nibblers.

Mothers nibble out of boredom or from diet deficiencies. Babies nibble because they are cutting their teeth. One year we raised seventeen colts on our ranch in California, and after all the babies had cut their teeth, there was scarcely a corral fence that hadn't been nibbled clear through.

There are new items on the market that you can coat your fences with to prevent this sort of mayhem, and you really should look into them if you want to keep your mare's boudoir from acquiring an air of delapidation with the passage of time.

A WORD ABOUT PORTABLE SHELTERS

The new portable corrals made of metal tubing quite neatly solve the housing problem as well as the problem of nibbling, so you should investigate their possibilities.

These new mobile shelters may be erected with a minimum amount of sweat and strain, taken down and moved about without too much effort, and they are extremely durable. They lack the romantic appeal of whitewashed or worm fences, and many horse lovers frown on them, considering their utilitarianism rather ugly. However, if money looms large as a problem, and you want a modern, convenient, easy-to-care for corral, you really should think about these metal mobile units.

Same orphan—fat and fresh—four months later.

COPING WITH CLIMATE PROBLEMS

Horses can stand a great deal of cold if they have room to move about in to keep warm. But they can stand very little cold if they are confined to quarters that do not permit extensive movement.

Also horses, like humans, can stand dry cold better than damp cold. If quarters are kept well bedded and dry, the scantiest sort of shelter may prove sufficient for your mare and foal. But if you allow your shelter to become over-damp both mother and foal will suffer and you will find yourself paying vet bills for distemper, thrush (a disease of the hoof caused by excessive dampness), and your foal will not stand too good a chance of survival.

Plan your shelter with an experienced horseman who knows enough about your particular weather conditions to provide mare and foal with shelter from prevailing wind, rain, and snow.

SOME AESTHETIC CONSIDERATIONS

If you are fortunate enough to be able to afford a genuine stable, however modest, see that sunshine reaches all parts of the mare's stall at one time or another during the day and be sure to include a window with a view, one that allows proper air circulation yet eliminates drafts.

In case the suggestion of a window with a view sounds a bit precious, remember your mare requires the stimulus of some form of interest besides four walls. She would be just as unhappy looking at nothing as you would be.

In developing a stall in which the mare may have her foal (though it would be much better for her to have it in a sheltered open space outdoors where there would be more room), be sure to allow sufficient area for the mare to lie down, stretch out, and have her foal without cramping either mother or babe. Too small quarters at birth can result in injury to mare or foal and should certainly be avoided.

ROOM TO SPARE

The mare should have room to spare in her quarters at

all times. She should be able to stretch out comfortably, and after the foal arrives, it should be able to stretch out just as comfortably with her. Also there should be room for the foal to romp about inside stable or stall if bad weather makes it impossible for it to be outdoors.

As the foal increases in size and gains strength, there should be ample outdoor space in which to exercise both the mare and her offspring.

WATER AND FEED

Make certain your "mare-ternity" ward has ample fresh water available at all times and that the water is kept clean. Sometimes an animal or bird may drop into a water dispenser and remain there unseen unless you inspect daily; and horses have an embarrassing habit of dumping stool in their water container and this too may go unnoticed unless you make daily inspections a habit. Water should not be allowed to become stagnant, and water containers should be emptied, scrubbed with a stiff wire brush, and aired regularly.

A feeding crib or manger should be provided for your mother-to-be but we will cover these important matters in greater depth later.

For the present we'd like to call to your attention a few additional matters that should be taken care of when you finally arrive at the moment of decision and claim one particular mare as your own.

What a thrilling decision it is!

How you love this velvet-nosed creature with its beguiling eyes, hay-scented breath, and heart-tugging whinny. What dreams your imagination conjures up as you look at the future mother of your foal and behold her ripe with promise!

Junior connects with his milk supply. (Courtesy Margie Moss, Canoga Park, California.)

But lofty as your aspirations may be, you must come down to earth long enough to attend to countless practical matters.

VET-CHECKING YOUR MARE

Before you exchange any money for your mare, you have a right to ask the owner's permission to have her vet-checked for soundness and general good health since you certainly don't want any mare bred unless she is in good health and capable of producing a healthy foal.

Besides this general check up, you will want the vet to guarantee that she is "open," that is, ready for breeding, with all internal organs in good shape and in normal position.

SOME ADDITIONAL PRECAUTIONS

The mare's teeth should be checked very thoroughly and "floated"—by that we mean filed to smoothness—if there are any rough edges that may cut her gums or interfere with proper mastication of her food.

A mare must be able to eat properly if she is to remain well nourished during her gestation period. And she can't eat properly unless her teeth are in good condition.

As horses get older they frequently develop hooks on their back teeth that can cause painful sores and also cause malnutrition. So be sure your mare's teeth have a vet's okay before you breed her.

A mare's feet should also be checked for founder, thrush, stone bruises, or contraction at the heel.

If you plan to ride her, then she must be properly shod if she is to travel over any rough roads. If she is to be turned out, have her shoes pulled and her hoofs trimmed; and keep her hoofs clean and trimmed throughout her pregnancy.

REGARDING SHOTS

A mare should also be given shots against distemper, tetanus, and abortion—the latter only if your vet feels it necessary on the basis of a record of previous difficulty of this nature.

ABOUT TRANSPORTATION

When you have selected the mare of your choice, have attended to all the necessary health checks just men-

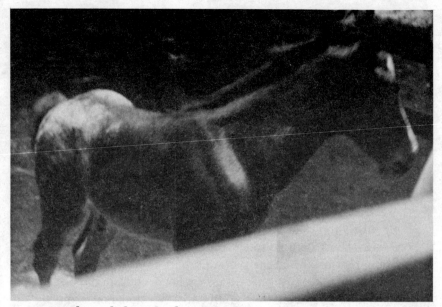

Appaloosa babies display spotted rumps that look as if they had been dipped in a paint bucket.

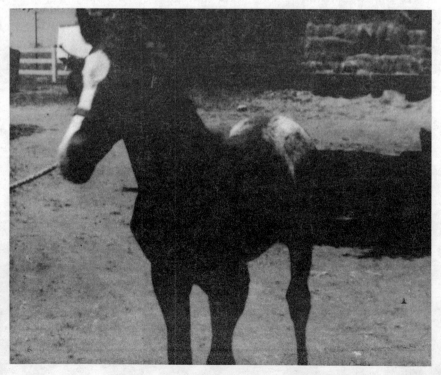

tioned, and have received a bill of sale proclaiming to all the world that you are her owner, you must next make arrangements to have her transported to your home by *insured* carrier. It is very important to choose a reliable person who is insured to transport your horses at any time. Never run the risk of dealing with irresponsible persons in the hope of saving a few dollars.

THINK TWICE ABOUT SO-CALLED BARGAINS

A poorly constructed trailer can cause injury to a valuable and beloved animal that cannot be reckoned in dollars and cents.

Be wary of people who offer to cart a horse for you at a discount rate. Usually their conveyances are roughshod, open trucks ill-suited to the purpose of transporting horses.

TRAILER TIPS

A horse trailer should have corrugated boarding on the bottom to provide a sure grip for your horse's hoofs, and preferably it should also be covered with rubber to prevent slipping. Sides should be well padded, and there should be proper arrangements for safe cross tying of any animals within the trailer.

Ventilation should be adequate but not drafty since your horse will frequently be warm from exercise when you transport her.

And most important of all, there should be a tail gate that provides a gently sloped ramp by means of which the mare may enter the trailer.

SKILL REQUIRED IN LOADING

Important as it may be to select the right trailer for your mare, you must be aware that it is just as important to choose someone with skill and patience in loading horses for the job of driving the mare into the trailer.

Some horses load quickly and easily if they are old-time veterans accustomed to being hauled about frequently to shows.

But many horses do not load readily. And all too often a fine horse has been ruined, injured, or rendered trailer-shy simply because of impatient or unskilful loading.

Since it is impossible for you to know the character or the ability of the person loading your horse till he loads it, you should shop around among knowledgeable horsemen, make careful inquiries, and find out in advance which persons may be entrusted with this task.

Or, better still, watch loaders at work and see for yourself how they load skittish, nervous, or difficult loaders. Anyone who grows angry, who shouts, or who flourishes a whip during the loading process is not for you.

A quiet-spoken horseman, using a rope skilfully, can usually load even a difficult loader with tact and patience —sometimes resorting to the enticement of a little grain. Such a person is certainly the only sort of person you would want to handle the mother of your precious foal-to-be.

WELCOME HOME

We hope, when your mare is loaded and headed for her new home, that you have made certain that her shelter, feed, water, gear, etc., are ready and waiting for her arrival. (We will go into greater detail about all these things later.)

You might have her name painted or burned into a decorative wood plaque and hang it on her stall or corral.

And you might have a few carrots and a dry bran mash, as well as some sweet feed, on hand as a token of friendship. We always give all new arrivals a bran mash as they

Coming three and already he gives his rider a relaxed, responsive ride, evidence of long months of training. (Courtesy Dee Cooper, Paramount Ranch, Agoura, California.)

almost always seem to be having some bowel problems, and this takes care of the situation.

A STRANGER IN A STRANGE PLACE

Remember that your mare will be ill at ease for several days after you first transport her to her new quarters.

She has been separated from people and horses she has known and loved and feels very much a lonesome stranger. Therefore she may not behave as she normally would at all.

Give her time to relax and to learn that she is among friends who will not harm her. Move quietly around her. Speak to her in gentle, soothing tones. Feed her some goodies out of your hand. No sugar, please. Sugar is bad for her teeth and for her disposition because if you withhold it once you have fed it to her, she becomes mean and snappish.

Love, love, love in copious quantities will soon win her over, and then you two can share together the wonder, the watching, and the waiting that lies ahead for both of you . . . all part of the pleasure of raising a colt!

3

CHOOSING A SUITABLE SIRE

THE FUN OF PICKING A PERFECT PAPA

Tracking down the perfect—or at least near perfect—papa for your colt-to-be can prove as much fun as your search for a model mother.

The first step, even before you start your quest, is to look through various breed journals.

If you find pictures of stallions that appeal to you you should drop a line to the breeding farm advertising the services of such stallions and arrange for an appointment so that you may visit the farm and see the stallion in person. Don't settle for a stallion on the basis of a picture alone because appearances may sometimes be deceiving, especially in a retouched photograph!

Take your camera when you visit the stud farm and also your Colt Raising Portfolio referred to when we were discussing your search for a mare; and be prepared to file away pictures and make endless notes before reaching a final decision.

As you did with your mare, take photos of any stallions that seem especially desirable and note any details of their background, progeny, coloring, conformation, performance achievements, etc.

A buckskin and a Palomino baby. They're led in tandem at eight weeks.

PAPERED OR NOT PAPERED PAPAS?

We decided in our discussion of brood mares that a mare need not always be papered; especially if you are breeding to one of the color lines such as an Appaloosa, Palomino, Albino, Pinto, etc., where stallions may be bred to grade mares and the offspring registered providing they show suitable breed characteristics.

But it is definitely a good idea to choose a papered, that is, a pedigreed, sire for the following reasons:

ADVANTAGES OF PEDIGREED SIRES

A papered or pedigreed stallion will usually produce get of finer conformation, better performance ability, and

CHOOSING A SUITABLE SIRE

greater stamina than a non-pedigreed stallion, not because such a stallion is worth more in a monetary sense—though this may be true—but because generations of excellent care and thoughtful selectivity enter into his breeding.

The offspring of a papered stallion will usually bring higher prices on the open market than the offspring of grade stallions and this is certainly a consideration to bear in mind if you are trying to make your colt raising project pay for itself.

The offspring of papered stallions when registered, may be entered in qualified horse shows, may engage in competition in such shows and thus become eligible for trophies, ribbons, etc., which not only adds to their owner's pleasure but increases their sales value on the basis of proven performance records.

SOME TIPS ON CHOOSING A SIRE

We have described the method to be followed in tracking down a satisfactory brood mare.

Relatively the same procedure may be followed in conducting a search for a suitable stallion.

You will want to visit several reputable breeding farms, make notes, take pictures, investigate performance records at shows and on the range. You will want to track down actual offspring wherever possible and see that they live up to the claims made for them by the stallion's owner.

Then, after thoughtful analysis of the data collected and careful comparison of all likely possibilities, choose a sire of the breed, color, disposition, conformation and general appearance that has the highest appeal to you.

SOME BREEDING PARTICULARS

When you have chosen a mate for your mare, you will

enter into a contract (which must be signed by someone of legal age) with the stallion's owner.

The contract will make demands on both you and the owner. As stated previously you will be expected to guarantee that your mare is in good health and "open" or ready for breeding.

You will be expected to pay for the mare's feed and lodging during the period of her stay at the breeding farm. This period may run as long as thirty days or longer depending on the amount of time required for the breeding to be verified.

If the mare should not "catch," that is become pregnant, first time round, she will have to remain for another three weeks until she comes in season again. Once bred you must wait at least three more weeks to see if she comes in season again. If she doesn't she may be said to be "caught" but only a vet check 45 to 90 days later will really serve to verify her pregnancy.

You will also be required to provide transportation to and from the stallion's headquarters.

BREEDER'S OBLIGATIONS

The breeder will be expected, depending on prearranged agreement, to provide a stallion in good health, hygienically clean and as many services (or breedings) as may be required to "settle" or impregnate the mare.

He should offer clean, well kept quarters for the mare with ample room for her foal, if she should be brought for breeding "with foal at side."

He will guarantee, this depending on the price of the breeding fee, a live foal or a live foal of a particular color.

A live foal is one that stands on its feet and nurses. If something happens to the foal later on the breeder is not obligated to supply a second breeding without charge.

The pleasure mare doubles the pleasure, doubles the fun if you breed her, for then you have a foal to look forward to. (Courtesy Vivian Cook, Agoura, California.)

At four weeks he submits to the lead rope.

Should the colt be stillborn, however, the breeder is expected to supply a second breeding at no additional charge.

Should a mare produce twins it is unusual indeed for both foals to survive. One foal will usually be too weak or too delicate to rear. In the case of Thoroughbreds, twin foals may not be registered.

But in the case of Appaloosas, where twin foals appear to be a breed characteristic and twin births are not at all uncommon, there are many authenticated records of both foals surviving and reaching maturity.

OTHER METHODS OF BREEDING

There are other methods of breeding that may be used to get your mare in foal, so you should be familiar with them.

A neighbor who owns a stallion may bring the stud to your ranch or farm and the mare may be bred on your property. This method of breeding is usually cheaper than the boarding method and may be quite satisfactory.

It will be necessary to make certain that the mare is in season, that she is ready for breeding, preferably in the second or third day of her heat, when the stallion arrives.

Experienced persons must handle mare and stallion to assure proper breeding with maximum safety for handlers and animals. This is no situation for an amateur since a stallion can be extremely dangerous if annoyed while his attention is directed to a mare in heat.

Any doubts about the mare's readiness for breeding may be dispelled simply by consulting a veterinarian.

Another method is simple and inexpensive and considered by many old school horsemen to be the safest and surest method of obtaining a guaranteed breeding. That is to turn your mare out at pasture with a stallion,

At four months the foal stands at command in halter.

of course first obtaining the owner's permission and making certain all fences are secure and that the two animals are compatible.

When the mare comes in season she will be bred by nature's own method with a minimum of fuss and inconvenience.

And of course there is always the possibility of breeding your mare through the process of artificial insemination. This method is still not too widely used and you would have to consult local vets, breed associations, experimental breeding farms, and agricultural colleges to find out just what could be done along these lines; but the investigation would be enlightening and under vet's supervision the results should be all you could hope they might be.

With any of these latter methods of breeding you will find it necessary to reach agreements on price, etc., through verbal negotiations with the stallion's owner or the person in charge of the experimental breeding plan.

If you wish to enter into a more formal sort of contract it will have to be worked out on a person-to-person basis with whatever papers seem legally suitable.

FINAL CHECK-UP

Forty-five days after breeding a vet check should reveal whether or not a pregnancy is under way. But many horse owners find it safer to wait ninety days when a vet check will more surely determine the presence or absence of a beginning foal.

You must not be too discouraged if: 1) a vet tells you your mare is in foal and she turns out not to be; 2) a vet tells you your mare is not in foal and she is; 3) your mare gives all the appearance of being in foal, even to making bag, then turns out not to be pregnant. Mares, like dogs, are sometimes given to false pregnancies, a very disappointing experience for the would-be breeder. But fortunately such mares are in the minority. So don't anticipate trouble of this sort as it is very rare.

In the vast majority of instances your mare will be "settled" first time round and you will start the long count of 336 days to delivery date, your mind and heart filled with all the wonder and promise and mystery that are such an exciting part of the fun of raising a colt.

4

THE LONG WAIT—WATCHING AND WONDERING

MAKING FRIENDS WITH MAMA

Your mare is yours at last and—hopefully—she is in foal. You have followed instructions carefully and prepared thoughtfully for her arrival.

Her stall is bedded down with clean straw, her manger filled with hay. Fresh, pure water is at hand, and you have made certain that she has ample room in which to stroll about and get exercise.

You have attended to your mare's teeth and feet, arranged for any shots she may have required, and had her wormed *before* breeding. Worming should never be done immediately after breeding since the mare may react unfavorably to the purging effect of the worm medicine.

You have assembled whatever gear, tack, and medical or grooming aids may be required for her care. And now comes the fun part when you begin at last to make friends with mama.

MAKE HASTE SLOWLY

As in any friendship, advances must be made slowly with due respect for the temperament, emotions, and likes and dislikes of both parties.

Stable mates enjoy a "window with a view."

Youngsters learn to eat together politely.

If you are patient enough to just watch your mare quietly for a while, letting her become accustomed to your presence and allowing her to make the first overtures toward you, your patience will be well rewarded.

Extend a hand with a little sweet feed in it and before long curiosity and hunger will overcome both shyness and fear, and she will soon be eating out of your hand. Thus a friendship begins that may mellow over the years into one of the happiest and most memorable relationships of your life.

THE PLEASURE OF YOUR COMPANY

Form the habit of patting your mare gently as she eats. Sit quietly by her, observing her motions and her mannerisms. The movement of her ears, the tensing of her muscles will tell you things about her: what she is interested in, what she fears.

Soon she will become used to having you nearby and will actually begin to enjoy the pleasure of your presence. Eventually she will relax and accept you as a part of her life that symbolizes security and contentment.

After following this procedure for a day or two you are ready to make further advances. These might consist of haltering the mare, leading her on a rope, and, assuming she is sufficiently gentle, cross tying her to two secure posts. We make a point of having the posts secure because if a horse should become frightened and pull back suddenly any post used for tying must be so firmly anchored that it won't give way and take off with the mare. If this happens the result could be a ruined animal.

We recommend cross tying because it is the safest method of tying any horse, especially when you intend to work around it. Older or very gentle horses or horses accustomed to ground tying do not present any problem

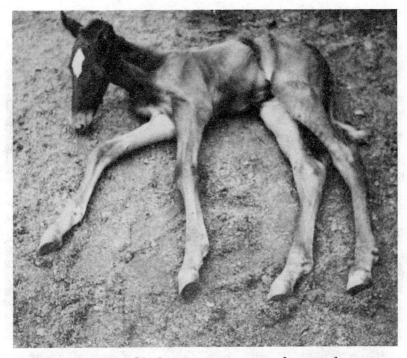

Only 15 minutes old, this youngster is just about ready to get to his feet. (Courtesy Linda Barnes, Torrance, California.)

in handling, but it is always well to assume a horse is not gentle till it proves that it is. And it is always well to work around any horse as if it were not gentle, not only to avoid unpleasant surprises but to accustom yourself to the safe method of handling a skittish or unmanageable horse when you face such an emergency.

TIPS ON CORRECT METHOD OF CROSS TYING

To cross tie properly you attach a tie rope to either

Four two-year-old Appaloosas display their colorful rumps. Blanket types and Leopards are represented. Who wouldn't love to produce colts like these? (Courtesy Dee Cooper, Paramount Ranch, Agoura, California.)

side of the halter and then anchor each rope, by means of a slip knot—never a hard knot—to any secure support.

We don't recommend tying any horse, however gentle, to fences, rickety corral rails, or just any object that comes to hand.

When your mare appears at ease and doesn't attempt to pull back or break away, you may take the next step

THE LONG WAIT

forward in your relationship—the pleasant task of grooming her.

A DISCUSSION OF GROOMING AIDS

To groom her properly you will need the items listed below. But you don't have to buy everything at once; get the most needed accessories first and add others gradually.

You will want a Bakelite rubber or a metal curry comb. The heavier metal comb is used when coats are heavy during the winter and dirt is stubbornly embedded.

For warm weather you'll want a non-toxic fly spray that won't harm her coat.

You'll also need a coarse brush for heavy grooming and a dandy or polishing brush for finishing touches.

When it is warm enough to hose her down, you'll want a metal scraper to remove excess water from her coat.

You'll need a hoof pick and a hoof dressing for grooming her hoofs.

A soft piece of flannel will come in handy for wiping out ears and nostrils and also for polishing her coat after brushing.

CORRECT METHOD OF GROOMING

To groom your mare properly start at her neck, work down her back and sides and under her stomach, using the curry comb to loosen dirt and the brush to remove the dirt once it is loosened.

Do not do her head or legs till you have completely groomed her body.

Stand facing the rear of the horse and always be on the alert to move quickly away from kicks or stompings. To anticipate sudden movements, watch the mare's muscles to see if she is tensing them.

In hot weather you will want to spray against flies before grooming, but before you use the spray be sure your mare is accustomed to being sprayed and won't shy from it.

POINTERS FOR EFFECTIVE GROOMING

As you grow more experienced at grooming, you may work with a curry comb and a brush simultaneously; but as a beginner you will probably want to use the curry comb first and then go over the mare's coat with a brush.

The effectiveness of the grooming may be determined by the amount of perspiration you work up while performing this chore. Unless you find yourself lathered with a good earthy sweat, you can't really say you have groomed your mare properly.

After currying and brushing with the cleaning brush, you will want to go over her coat with the dandy or finishing brush.

Then, depending on your enthusiasm or stamina, you may add polishing with the piece of clean flannel, and finally, you'll add the supreme touch—truly a labor of love—going over the entire coat with the palms of your hands. The real horse lover insists this is the only way to bring out that lasting lustrous sheen that is such a joy to behold.

If weather conditions are favorable and your mare is accustomed to being hosed down, you may wish to give her a hosing and then run the scraper over her. Allow her to dry thoroughly and keep her tied till she has no moisture left on her coat or she will most certainly roll in the nearest dust pile, undoing all your fine efforts. As soon as his coat becomes damp, a horse likes nothing better than to roll in dust—apparently considering this equivalent to a dusting with talcum powder. So keep this

The wistful eyes of brood mares in the auction lot say louder than words: "Please take me to a good home where I can have my foal in happy surroundings."

idiosyncracy in mind if you wish to save yourself lots of heartache.

We do not recommend bathing your mare in the later stages of pregnancy since dirty water may run down over the mare's bag, which would not be too good for her.

FURTHER NICETIES OF GROOMING

After grooming her neck and body you are ready to tackle her head, legs, mane, and tail, in that order.

Her head and legs are sensitive, so she will not tolerate the use of a metal curry comb or too rough a brush on these areas.

Also she may be head shy from previous improper handling, so you should approach the grooming of her head with great care.

Ears and nostrils especially are highly sensitive and should be handled accordingly.

THE GENTLE TOUCH PAYS OFF

All movements during grooming should be slow-paced and gentle. If you talk to the mare, and this may be an excellent way to quiet her, speak softly in reassuring tones. Soon this little interchange of verbal commentary will become one of the most pleasurable aspects of the grooming period for both of you.

Each grooming implement should be shown to the mare before it is used, and she should be allowed to sniff it if possible. In this way she will become familiar with it and accept it before you use it on her.

Her head may be brushed softly from the poll downward with a very soft brush or a piece of flannel.

Ears may be wiped out with a clean, soft cloth, but you must work very carefully about the ears and not pull on them or bend them. The expert horseman will tell you

THE LONG WAIT

a horse's ears should always be handled as delicately as you would hold a bird in your hand.

Eyes may be wiped out with a small sponge that has been dipped in clear water and wrung out. But be sure the mare is aware of your purpose and consents to it or you may face a serious pull back.

Nostrils should also be wiped out ever so tenderly with a damp cloth as dust collects in them when the mare eats her hay.

BOTTS CAN BE BOTHERSOME

The bott, a small fly that resembles a bee in appearance but is in no way related to the bee family, can give your mare a troublesome time during bott season, usually the latter part of the summer.

Horses seem to have an instinctive fear of botts. When you see horses tearing madly around at pasture for no apparent reason, snorting, sliding to a halt, and taking off again as if the devil himself were after them, you can almost be certain that botts are giving them a bad time.

If you observe a bott's action around the horse, you will be even more convinced that a bee is at work for the bott has an ovipository, or downward hanging appendage, with which it deposits its eggs, one on each hair of the horse; and as it does so you would swear it is stinging the animal, but it isn't.

Bott eggs usually are laid on the ankles or shoulders of the horse, where the bott knows they will most certainly be nipped at and swallowed.

If swallowed the eggs produce worms that are injurious to the horse and can cause serious trouble. We have found the safest way to remove the bott eggs is to go over the horse's hide, wherever the eggs have been deposited, and remove each egg with a straight-edge razor blade. It

66   THE FUN OF RAISING A COLT

Mama tells baby a few facts of life.

doesn't hurt the horse at all, but it effectively disposes of the eggs. During bott season we sometimes have to perform this task every day.

GROOMING MANES AND TAILS

Manes and tails should be washed with a sponge dipped in water made sudsy with liquid dishwashing detergent. We sometimes add water softener to increase cleansing action and to leave the hair pliable. In the case of silver manes and tails, a little bleach may be added to the rinse water, but use minimum quantities. We use about three tablespoonful of bleach to half a bucket of water.

Special shampoos are available for grooming manes and tails, and you may wish to investigate them.

A good way to do the tail is to hold a bucket filled with your sudsy water under the tail, allow the tail to drop into the bucket, and keep running your hand down the tail from top to bottom, squeezing out the moisture with your fingers.

Only a gentle horse will permit this of course, and you should stand to one side as you work, keeping alert for hoof movements.

Suds should be rinsed out thoroughly with clear water and the rinsing continued till the water is no longer sudsy.

ELIMINATION OF TANGLES

Manes and tails may be combed with a special comb made for this purpose. You will find the hair much easier to comb after it is wet than when it is dry.

Burrs and mattings should be removed promptly, especially when the fly season is on, and the mare switches her tail continually. If tangles are allowed to accumulate, she may switch her tail into a hard ball that you will be unable to disentangle. Pretty soon she won't have any tail, for you will have to cut out the mattings, leaving her none.

When manes and tails are dry they may be brushed. Of course you will also brush the forelock. When you want to be really fancy you will braid your mare's mane and tail, and tie with ribbon to encourage waves—simply as a pleasant, glamorous touch.

Some horse lovers attempt to brush their horse's teeth,

but we never have done this. We do inspect teeth and gums regularly to be sure tartar isn't collecting and foxtails and bits of wire or wood aren't embedded in the gums.

REGARDING HOOF CARE

We recommend that you pay particular attention to the regular care of your mare's hoofs. Lift each one daily to see that there are no stones or nails present and remove all accumulations of mud or caked straw with a hoof pick. If you don't do this you risk the danger of thrush, a hoof ailment caused by unclean, damp stalls or corrals.

See illustrations for correct method of handling both front and back hoofs in grooming.

After cleaning, hoofs may be dressed with a special hoof conditioner or brushed with crank case oil obtained from your nearest service station. This treatment tends to keep hoofs from becoming brittle or cracking.

A WORD ABOUT SHOEING

Your mare should be kept shod if she is being ridden up to the last weeks of pregnancy, at which time her shoes may be pulled to give her feet a rest. She should never be ridden unshod over any hard or stony surface.

Shoes should be changed about every six weeks, depending on the rate of hoof growth, to avoid contraction of the heel. If the mare is kept unshod, then her hoofs should be trimmed regularly, before they crack into the quick.

It is important for you to locate a capable and trustworthy blacksmith as proper shoeing can add so much to the pleasure of your riding and the good health of your mare. Her feet should not be neglected any more than you would neglect your own feet. It will be up to

THE LONG WAIT 69

Broodmare heavy in foal dreams away the hours till her colt is due.

the blacksmith to tell you whether hot or cold shoeing is required. Hot shoeing is more expensive, but we have always found it more satisfactory, though many horsemen find nothing wrong with the cold shoeing method.

A wise and experienced blacksmith will tell you much about horse lore, both fact and fancy, that will enlarge your knowledge of horses and horse care. Listen and learn. You will make a friend and you will find the happy times of shoeing, with the pungent scent of the brazier, the sharp clang of the iron on the anvil, and the pleasant conversation, part of your pleasantest memories of colt raising.

We always look forward to the visits of the blacksmith to our ranch, the Bar None. Even the dogs rejoice for they like nothing better than to nibble on the hoof parings that lie scattered about at the end of the shoeing. It seems these tidbits supply something essential to the dog's diet as well as good exercise in chewing.

On a warm summer day, as you sit beneath the shade of a spreading tree and chat with your blacksmith, watching your mare switch her tail lazily, you'll understand the deep and lasting bond that exists between all horse lovers.

By now you have established a routine of daily loving care, the importance of which cannot be stressed too highly. This personal, solicitous contact between you and your mare cements your friendship as nothing else can.

T.L.C.—TENDER LOVING CARE—IN DAILY DOSES

It is these daily doses of tender loving care in generous quantities that soothe your mare's nerves, sweeten her disposition, cause her to feel relaxed and cherished, and most certainly influence her health and the health of her colt.

You can't spend too much time with your mare. You will see, as your friendship deepens, that she soon looks forward to these daily love-ins, associates her pleasure with you, and eventually welcomes you into her heart.

Besides the closeness and pleasure of grooming, you

THE LONG WAIT

will want to sit quietly with your mare while she eats or grazes nearby.

Together you will dream of the foal-to-be. She in placid contentment, you in awesome wonder at the silent miracle of impending birth.

We've always believed that the mare having her first foal has no awareness at all of what is going on within her. She may imagine that she has an extra-full tummy as the colt grows in size. And in the latter stages of pregnancy, she may feel mighty uncomfortable, especially when her four legged companion stretches its limbs and indulges in a few kicks. But until the foal actually arrives, it is doubtful if the mare anticipates motherhood.

KEEPING A CALENDAR AND NOTING DOWN CHANGES

You will find that it's fun to keep a calendar on which you can X-out each passing day slowly, oh ever so slowly, working your way to the magic 336, which may or may not be delivery day.

Changes will occur in your mare's appearance before your eyes. Gradually her belly will swell and deepen. You will be certain you can see her flanks move as the colt kicks within her. You will rest your hand against her flank and swear that you feel the actual movement of the unborn foal. (We engaged in this happy practice for a full twelve months with one mare we owned only to discover that she wasn't in foal!)

All this while you will make certain that she receives ample hay and water and gets plenty of healthful exercise.

TO RIDE OR NOT TO RIDE IS OFTEN A QUESTION

You can ride your mare and, in fact, should ride her, if she has been broken to ride, throughout most of her pregnancy.

She should not be ridden hard or allowed to become winded or overheated, but you can ride her at a gentle jog or a leisurely walk right up to the time she starts to "make bag"—usually within four to six weeks of delivery.

It would not be wise to use a heavy or over-tight saddle on her. Instead use a lightrider (surcingle pad and stirrups) or ride her bareback. If you must use a saddle, make it a light one and girth up just sufficiently to hold it in place, not enough to distress her.

In the last month or two your vet may suggest special supplementary feeds for your mare's diet—tonics, vitamin boosters, or general conditioners. Don't over-feed or give her too much protein, but see that her hunger is satisfied. Your vet will have good advice for you on this score.

Just be sure she always has plenty of water, and we mean copious quantities. As a milk-maker she needs them.

Grooming, exercising, watching, and loving, through almost a full year of days, all add more and more to the treasure-filled memories and pleasure-filled days that are part of the pure fun of raising a colt.

5

AS THE BIG DAY APPROACHES

TIME DRAGS FOR LADIES IN WAITING

You are certain, when you start your long vigil waiting for your colt to be born, that it will be impossible to contain yourself and to wait patiently for the 336 days to grind by at snail's pace as you and your mare approach the big day of delivery.

But you must remember that you alone are the impatient one.

The mare knows nothing of the gestation process going on within her and, as we mentioned previously, she has no sense of the passage of time, no feeling either of dread or of anticipation with regard to the future.

As the foal grows larger she may experience a certain amount of discomfort, but scarcely more than that which might be occasioned by a mild case of colic.

When the foal moves, turns over restively, or kicks within her, she may be mildly surprised, but she will not be stirred emotionally.

As long as she is loved, cared for, fed and watered, and enjoys a moderate amount of exercise, she will remain oblivious of any portending event right up to the start of her labor pains.

But you, alas, will suffer prodigiously, vicariously,

vociferously—that is, you will if you're like most horse lovers. We suffer the same agonies through every gestation period, this in spite of the fact that we've travailed through so many we can't count 'em, and one year raised 17 foals on our ranch!

MAKING BAG AND OTHER SIGNS OF IMMINENCY

From the time the mare starts to "make bag," that is, from the time that her bag begins noticeably to increase in size, you may assume that delivery is probable if not imminent.

But oh how slowly her milk accumulates, depending of course on the individual mare, the amount of feed, water, grass, or diet supplements she is enjoying. There are so many factors to accelerate or diminish milk production that yau can't really be sure if the mare is entering on the last four to six weeks of pregnancy or not.

However, this is usually the time when her bag will begin to enlarge. Of course if you are dealing with a false pregnancy, bag enlargement will not be a reliable sign of impending birth. But we will assume your mare is really pregnant and figure accordingly.

Still there is so much variation in shapes and sizes of milk bags that you never really know exactly what's up. Some mares have most of their bag concealed within the body cavity and show very little evidence of enlargement right up to the time of delivery. This is frequently true of mares having a first foal.

A LITTLE STRANGER MAY ARRIVE AS A BIG SURPRISE

If a mare with a concealed milk bag has been bred accidentally at pasture without your knowledge, you may be in for a big surprise, for she may present you with a little stranger one fine morning when you haven't even dreamed that she might be pregnant.

AS THE BIG DAY APPROACHES

Brood mares play together . . .

Actually this is a very pleasant way of becoming the proud owner of a foal without any of the sweat, suffering, or suspense usually associated with the ordinary gestation period. It is especially merciful to be spared those last endless weeks of waiting when time seems to move backward rather than forward.

MONKEY BUSINESS IN THE OLD CORRAL

As we write this we call to mind just such an incident, which we would like to share with you because it makes our point and we believe you will find it amusing as we did.

... and eat together while waiting for D (Delivery) Day.

AS THE BIG DAY APPROACHES

We sold a young, two-year-old filly to a dear friend and about eleven months later the friend called one evening to ask if we knew that the filly was in foal when we sold her. We said we most certainly did not since we considered the filly too young to breed. Our friend said something funny must be going on because the youngster was making bag.

We advised a vet check and our friend called back after the vet's visit to inform us her filly was due to produce a foal within a couple of weeks.

We recalled that the day before we had trailered the filly to our friend's ranch our neighbor's stud had turned up in our pasture. But we had led him home without giving a second thought to the matter.

Within two weeks our friend called to say the little filly had produced a handsome Appaloosa stud colt. Certainly one couldn't have a happier experience than to discover a little loud-colored Appaloosa in your corral one morning without ever suspecting it was on its way.

WILL YOU BE LUCKY ENOUGH TO WITNESS THE GRAND EVENT?

Most waiting periods are far more strenuous than the one just recounted, and it is seldom that you are priviledged to witness the actual birthing.

We have watched and waited through that last breathless week before delivery, scarcely letting the mare out of our sight night or day only to have a foal appear while we were indoors snatching a much-needed cup of coffee.

We've sneaked back of barns, peered from behind trees, spied from beneath corral rails only to have a mare refuse to have her foal for lack of privacy. Mares, it seems, can actually hold back delivery of a foal if their surroundings are not peaceful or if there is any element of danger threatening. Perhaps this is a protective mechanism bestowed on them by nature that preserves them in their wild state.

But when we have been quiet enough and patient enough, we have had mares considerate enough to cooperate with us and produce a foal right before our eyes—some at night, others in broad daylight. And what a rewarding experience this is. More than worth all the long days of stomach-tightening tension and suspense!

WHEN IS THE MARE SERIOUS ABOUT FOALING?

How will you know when your mare is really getting serious about having that foal?

Here are some signs to guide you:

Her bag will become greatly distended and milk veins will show over the length of her belly, possibly clear to her chest.

Within three or four days of foaling, the ends of her teats will exude a wax-like deposit that will cling to them in clear, amber-colored droplets.

Twenty-four to thirty-six hours before foaling, if a mare is a plenteous milk-maker, she will usually have so much milk in her bag that milk will spurt from it whenever she moves.

ACTIVE TO THE FINISH LINE

It is necessary to keep a mare active during the latter days of her pregnancy. If she tends to stand around motionless from discomfort, as some mares do, her legs may swell or she may become "stocked up," as horsemen say.

Watch for symptoms of this sort and lead her around on the end of a longue line for an hour or so each day.

LAST-MINUTE SIGNALS

When the mare begins to slacken off on her feed, she

Like mother, like son. Here we see a chestnut mare with her chestnut foal enjoying an afternoon siesta in the shade of some live oaks.

may be quite close to foaling. Don't force any feed on her at this time. Light rations are in order, and it would be quite all right if she were to stop eating altogether.

If she becomes restless, paces back and forth, looks toward her flank nervously, switches her tail, and breaks out in a sweat, chances are she has already started labor. If you place your hand along her flank or against her belly you may be able to feel the contractions.

Her instinct at this time will be to seek privacy: to get away from light, noise, other animals and strangers.

RESPECT YOUR MARE'S PRIVACY

This is not the time to invite curious friends or well-meaning neighbors to view the proceedings. Grant your mare the courtesy and consideration you would accord any mother-to-be and allow her to withdraw from the public gaze.

She would like to be entirely alone but you have waited so long for this moment and you want so much to be part of it that you have a right to be present. And, if you have won her trust and affection through the long days of loving care that you have given her, she will make concessions for you and may quite possibly allow you to remain near her.

If she does please observe certain amenities.

Stand quietly out of sight and don't try to be helpful. Many a foal has been injured at birth through the too eager ministrations of pseudo-experts who thought they were being helpful when actually they were only causing trouble.

LET NATURE TAKE ITS COURSE

Nine times out of ten a mare delivers her foal promptly and without difficulty if she is in good condition, has been properly cared for and has roomy and clean quarters.

When it appears that the mare is ready to foal, if the weather is mild and dry, she will be happier outdoors.

If she is in a corral, you must be sure that the ground is not hard, rocky, or muddy and that at least part of the corral is deeply covered in clean straw.

The corral should be wide enough and the fencing so arranged that she can lie down and have her foal without

Mother looks over her new foal, just ten minutes old. (Courtesy Pat Gast, Dancing Bear Ranch, Canyon Country, California.)

A pony foal—most fun of all!

AS THE BIG DAY APPROACHES

entangling her legs or endangering the foal as it is born.

The same principle applies to a stall. It should be clean, dry, thickly bedded, and sufficiently long and wide for her to have her foal comfortably.

HAVE EXPERIENCED HELP STANDING BY OR ON CALL

An experienced horseman or vet should be standing by or on call when the mare appears to be in labor. A healthy mare delivers quickly, usually within twenty to thirty minutes, and without any need for human assistance.

If something goes wrong and assistance is required, there is no time to seek aid once delivery has started; and this is certainly no place for an amateur. If trouble develops mare and foal can only be saved by the prompt action of someone who knows exactly what needs to be done and who has the strength as well as the skill required so cope with the situation.

AT THE ACTUAL TIME OF BIRTH

When the mare's water bag breaks, the foal is not long in arriving. The mare should lie down when this happens. If she should remain standing when she delivers the foal it may be injured if not killed.

If all is proceeding normally the front feet of the foal should present themselves, hoofs pointed *downward*.

If the underside of the hoof shows, the foal is in a breech position and a vet's assistance will be needed immediately.

If only one hoof presents itself in a downward position, it may mean the other leg is doubled back so that the hoofs cannot come forth normally.

Skilled hands will have to reach in at a time like this, and gently free the hooked leg if the birth is to proceed in a normal manner. No inexperienced person should proffer assistance as any wrong moves could cause serious damage to both mare and foal.

SIGNS OF NORMAL DELIVERY

Hoofs together, underside *down*, should be followed by the foal's legs, head tucked neatly between them.

As the head emerges you must see that air is reaching the foal plentifully. If the amniotic sac has not been opened, it is a good idea to open it and draw it back from the foal's nostrils, checking at the same time to see that the nostrils and mouth are free of mucus and removing any accumulations of mucus that may be present.

The widest part of the foal, the shoulders, should wriggle forth next, first one, then the other. This is the most difficult period of the delivery for the mare and is equivalent to the presentation of the head in a human birth. If assistance is needed, it may be given in the form of a slight pull forward on the foal's legs. The pull must be rendered expertly and gently, but with sufficient strength behind it to really aid in the delivery of the foal. The pull must be timed to coincide precisely with the mare's expelling contractions; otherwise mare and foal could be seriously injured.

The mare should be allowed to rest quietly between contractions and the foal should be allowed to rest also.

With the final expulsion the foal will be free but the umbilical cord will still be attached to the placenta of the mare; neither mare nor foal should be disturbed at this juncture.

WHEN THE FOAL TAKES ITS FIRST STEPS

In due time and at its own pace, usually within 15 to 20 minutes, depending on its size and strength, the foal will struggle erect and stagger to its feet. In doing so it

A young foal enjoys a siesta.

Laticia Gast and her dad Gene being introduced to the new foal, Foggy Bear. (Courtesy Pat Gast, Dancing Bear Ranch, Canyon Country, California.)

AS THE BIG DAY APPROACHES

will usually break the umbilical cord in a natural way; and that should be the end of the matter.

If you wish to apply an antiseptic, by all means do so. If an expert is present, let him take whatever steps seem advisable. We leave our mares and foals strictly alone at this time and have never had any trouble. But this is a matter of individual choice and we suggest that you follow your vet's advice.

SHE'S UP, SHE'S DOWN, SHE'S DOWN, SHE'S UP

Either mother or foal may rise first—it makes no difference. A weak or delicate foal may attempt to get to its feet, almost make it, then sink to the ground several times before it finally manages to control its gangling legs. This is no cause for alarm. The foal will gather strength quickly if left to its own devices, though a really backward foal may need some assistance.

FILLY OR STUD, THE FIRST QUESTION ON ARRIVAL

Is it a girl or is it a boy?

This is the first question that pops into the mind of every horse lover as the newborn foal gets to its feet. Many are so excited at a time like this that they mistake one sex for another, usually imagining the foal is of the sex they most desire.

We've had this happen to us. And just this week in our local paper we noted in the horse club column that two owners of new foals had to make public announcements correcting statements made earlier regarding the sex of their new arrivals.

Unless you're in the breeding business you'll probably want a filly rather than a stud since the filly can always be bred whereas the stud, unless you wish to keep him for breeding purposes, will eventually have to be gelded, which costs quite a bit of money.

SOME FACTS ABOUT THE AFTERBIRTH

When the mare finally gets to her feet, she will, in all likelihood, be trailing portions of the afterbirth from the birth canal. The membraneous amniotic sac in which the foal emerged from its mother and which looks so much like silver parachute silk, has been removed from the foal by now, either by its mother or with your help. All this waste material should be gathered and burned so as not to attract flies or wild animals.

The afterbirth may depend from the vaginal canal for as long as 24 hours. Opinion differs among experts as to whether or not this should be tampered with. We suggest that you follow your vet's advice, but we will tell you how we cope with the situation.

We allow the mare to trail the placenta till it disengages itself normally even though she may step on it now and then or the foal may. We believe this is normal procedure in a mare in her wild state and we try always, in caring for our horses, to simulate as closely as possible the horse's natural conditions.

Some say it may injure the mare if she steps on the placenta and they tie it up out of reach of her heels. Others say that by stepping on the placenta the mare loosens and removes it promptly and completely.

We have no desire to take sides in this endless controversy but will merely state that if the mare doesn't "come clean"—that is, if she doesn't expell the placenta within 24 hours after giving birth to her foal—you should call a vet to do the cleaning job. Failure to do this can result in a barren mare, so it is only good sense to call on professional help if it is needed.

FIRST FEED AFTER FOALING

We always make it a point to give our mares a hot

This pinto filly at three repays long years of time and effort by appearing so regal and beautiful.

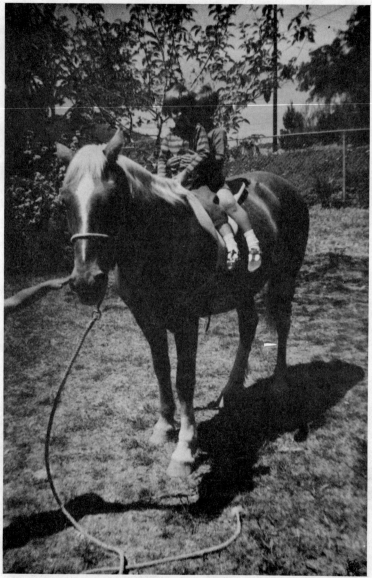

Perfect companions: a gentle mare and her two young owners. When the young colt arrives, who will enjoy it most, its mother or its foster mothers? (Courtesy Linda Ernst, Agoura, California.)

Boy and colt learn discipline as the young horse in training is led in harness.

WORK IN A CONTAINED AREA WITH AN ASSISTANT

It is well to work in a contained area until the foal leads without too much reluctance, and it is well to have an assistant, preferably an experienced horseman, close by so that if the foal should break away he can be caught and returned to you without your having to dismount.

If the assistant walks close behind and to one side of the foal as you start off, the foal will stay in place. With a western saddle you can take a half hitch around the horn with the end of the lead rope and thus keep things a little more under control.

If you ride English, however, or if you have no assistant to hand you the lead rope, you will have to position the foal close to the right or "off" side of the mare, slip the rope over the mare's withers just in front of the saddle, and secure it there with your left hand, along with the reins, as you mount.

ANOTHER STEP FORWARD IN DISCIPLINE

When the foal has learned to lead obediently alongside his mother, you will usually find that he also leads far more tractably from the ground. This means you can now take him forward to the next step of discipline.

To do this attach a longue rein to his mother's halter and longue or as some horsemen say, lunge (or drive), his mother in a circle and let the foal follow at will.

Before long he will begin to make the circle with her and when he does this with some degree of control and regularity he is ready to be longued on a line of his own.

INTRODUCING THE SURCINGLE

Next you are ready to put a surcingle on your colt. Tie him in a stall and be sure his butt is away from you and positioned in a corner before you attempt this. Introduce

Breaking the colt to drive in harness usually results in a gentler horse less apt to be kicky.

the surcingle slowly, allowing him to study it and sniff it before you try to put it on him. This procedure must be followed with any new object you wish to bring to the foal's attention.

After he has become familiar with the surcingle, you

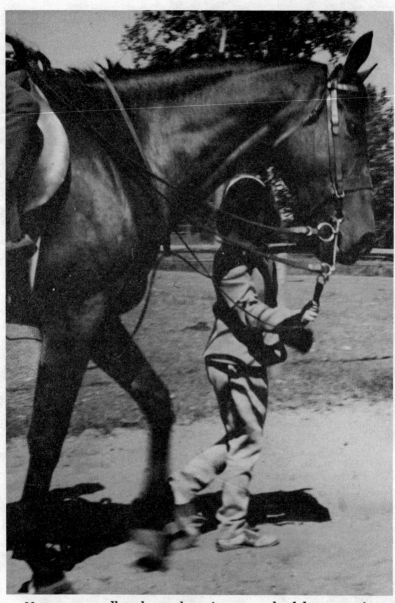

Never too small to learn the safe way to lead her mare for daily exercise. (Courtesy June Stenson.)

may lower it very gently over his back and allow it to rest there lightly, unfastened. Do this several times in succession, a few minutes at a time, before you make any effort to tighten it. Then draw it up enough to fasten it securely but not tight enough to cause the foal to buck, pull back, or rear. However, it must be fastened securely enough not to slip around under his belly for this could lead to an even greater catastrophe from which it might take weeks for him to recover.

INTRODUCTION OF THE PLAY BIT

Later you can introduce your pupil to a pony bit attached to an overcheck and let him play with this while he is in a stall or on the end of the longue line.

At first, however, attach a piece of smooth metal, possibly a key, at the hinge of the bit, making sure it will not come off to be swallowed and that it will be gentle on his tongue. Now he has his very own play bit or pacifier.

The foal will toy with this bit by the hour and it will work wonders for him, moistening his mouth, hardening his gums, keeping his neck supple. And of course all this while the foal becomes accustomed to the feel of a bit in his mouth. No harsh pressure is exerted, just enough to let him know the bit is present and to encourage him to work it.

Trotting beside his mother, exercising in a circle on the longue line, working with a toy bit, and being led about and groomed give the foal plenty of healthful muscle-building and character-building activity.

Now he is ready to leave his foalhood behind him and to go forward to colthood.

By this time he should stand when he is tied without pulling back; he should follow along on a lead rope without bucking or attempting to break away; and he should

Too small for her britches but with a heart big enough to know the thrill of affection for her very own horse. (Courtesy June Stenson.)

EARLY DISCIPLINE

Baby is introduced to hay and watches with eager curiosity as mother chomps contentedly.

accept the restraints and disciplines that go with human companionship with as much grace as he accepts human attention and affection.

REWARDS AS INCENTIVES

One way to make sure that he does this is to reward every display of good behavior—not with a carrot, an apple, or some sweet grain alone, but with a pat of approval and a word of praise.

Training authorities agree that rewards for obedience in training animals are much more effective than scoldings or punishment for disobedience.

You will find a foal very sensitive to the tone of your voice and very responsive to the cadence which tells him he is pleasing you. In all situations and under all circumstances try to remain calm, patient, and free from anger while you are disciplining him. Anger, impatience, cruelty of any kind—such as beating—or the use of violent coercion or restraint may win temporary obedience through fear, but the battle for ultimate control will be lost.

Lose your temper and you lose the possibility of becoming a respected professional horse trainer. You are not only teaching your foal manners, obedience, and performance skills, you are in the process of building his character. Remember that in any contest of wills with a horse, however young or old, you will invariably prove the loser. And certainly you can never hope to win any contest of strength.

Therefore the amount of patience, gentleness, and, above all, love that you express will determine your success as well as your pleasure in raising a colt.

9

MORE ADVANCED TRAINING

SEVERING THE SILVER CORD

Inevitably, in any parent-child relationship there comes a point in time when interdependence must cease. The joys and responsibilities of motherhood must be terminated and the securities and guaranteed satisfactions of childhood must end.

What a difficult period this can be for human parents. And children. But how deftly the situation is handled in most instances by the mothers of the animal kingdom.

When a mare's instinct tells her it is time to wean her foal because her milk supply is drying up, because another foal is due, because her foal is becoming obstreperous and physically too strong to cope with, or simply because she has suddenly become bored with the limitations and restrictions of motherhood, if left to her own devices she will quickly put junior in his place and on his own.

JUNIOR ACCEPTS ADULTHOOD

She does this usually by means of certain well placed kicks (never hard enough to hurt) and certain well timed nips (frequently these can hurt mightily).

Junior, when sufficiently battered and bruised, even-

More advanced training begins with an attempt to teach the colt the discipline of the bit and tie down.

tually gets the idea that this comforting, mothering creature who has met all his needs and supplied him with food, companionship and protection now most definitely wants no part of him.

He may sulk and complain for a while and hang around

MORE ADVANCED TRAINING 151

wistfully hoping things will change and the old order be restored, but when a second foal appears or a new playmate his own age turns up, or he is engaged in training that occupies his mind, he soon leaves his foalhood behind and accepts his new status as colt.

WHEN WEANING MUST BE FORCED

Sometimes, of course, the transition from foalhood to

Jace Garcia of Thousand Oaks, California shows off his trick Appaloosa gelding, Amigo. Many hours of tireless work go into training of this sort.

152   THE FUN OF RAISING A COLT

colthood is not accomplished this easily. As with human parents, some mares are over-protective and too indulgent to accomplish effective weaning.

Also some foals like human mama's boys hang back and cling to mom's apron strings. In such cases you must step into the picture and, with a firm hand, bring about the necessary weaning.

This means that you must place the mare in an enclosure where she will be secured in such a way that she cannot possibly jump out or hurt herself. And you may believe us when we say that a mare who is determined to rejoin her foal can surmount incredible obstacles.

We have seen a mare in such a situation jump a nine foot fence *uphill* though she had never before jumped even the low board of a cavaletti.

The foal, too, must be placed *out of sight* and, if possible, out of the hearing of the mare within an enclosure from which he cannot possibly escape and where he cannot hurt himself. We must warn you that a foal can jump remarkable obstacles and wiggle through impossible openings when he is determined to rejoin his mother.

HARDEN YOUR HEART IF YOU WOULD SHORTEN THE WEANING PROCESS

Both the foal and the mare will whinny continuously and piteously day and night at their first separation. It is a heart-rending sound that will cause you to wince every time you hear it and you will long to allow the two desolate creatures to be reunited.

But you must resist this temptation and harden your heart against their dramatic dialogue. To allow even a temporary reunion would be to prolong the agony.

Hold your ground and in two or three days—or a week at most—the mare will have forgotten her foal, the foal its

Jim Angell of Vallejo, California shows correct method of leading a young foal.

mother. Don't be fooled, however, by this seeming capitulation.

Keep mare and foal separated for at least three more weeks and then don't let them pasture or stall together or you may have the whole weaning job to do over.

THE FUN OF RAISING A COLT

OUT OF SIGHT, OUT OF HEART

It is always astonishing to discover on allowing the mare and foal (now a colt) to be reunited after a long separation that they eye each other with cool indifference. Sometimes there may be a mild whinny of recognition.

Baby leads the way into trailer, mama follows; owner Margie Moss shows skill in handling.

MORE ADVANCED TRAINING 155

More often the cool indifference chills even further, as it should.

The mare has a new foal to raise or she has earned a rest from motherhood. And the foal must now devote himself to sterner lessons in obedience, as he begins advanced training.

Before we discuss the points of such training we must take up the matter of "drying up the mare."

POINTERS ON DRYING UP THE MARE

We should preface our remarks by stating that two horsemen seldom agree as to the approved method of drying up a mare. Most divide into two schools of thought.

The first school prefers to strip the mare, that is to milk her regularly till her milk diminishes to a negligible quantity.

The second school simply cuts off all feed save minimal amounts of hay, limits water intake to bare necessity, and leaves the mare's bag strictly alone. We belong to this latter school.

When handled this way the bag will become distended and hard (often alarming to the uninitiated) and may require daily greasing with vaseline, Aquaphor, Albolene, or any similar water repellent.

If this is your first attempt at this method of weaning you will be convinced that the mare's bag may burst, that the mare must be suffering intensely, and that something should be done to relieve her or dire consequences may result.

We can assure you that if let alone, after ten days or so the bag will begin to grow smaller and soon will shrivel to a normal flaccid appearance.

Some horsemen milk about half a cup of liquid from the bag after ten days. We don't. We just let nature take

Chest pad devised by Margie Moss of Agoura, California protects the young foal whose inquisitiveness has caused chest injury.

its course and the course, at least in our experience, has always been quite satisfactory, from the mare's point of view as well as from ours.

FURTHER PRECAUTIONS DURING THE DRYING UP PERIOD

The mare may be returned to a normal diet once her bag is free of milk. She should then be allowed to drink all the water she craves.

If in the course of the drying up process any untoward symptoms occur, if the mare becomes restless or feverish (you can usually tell if she has a fever by feeling the tips

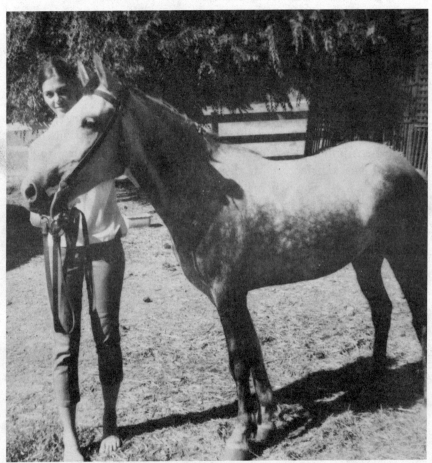

Young owner, Kay Lloyd of Agoura, California displays her colt, Kooskia, raised as a 4-H project.

of her ears which will feel hot to the touch), if she refuses to eat or walks awkwardly, then by all means call your vet and follow his advice.

Some veterinarians offer shots to hasten the drying, but we have never found such aids necessary.

AFTER MOTHERHOOD WHAT?

The mare is now ready to have her next foal if she has been bred back within thirty days after the birth of her previous foal or she is ready to be enjoyed once more as a pleasure mare.

If she has proved to be a good brood mare, bringing a foal to term and delivering it without complications, if she supplies plenty of nourishing milk, and if her offspring proves satisfactory in disposition, appearance, and performance, you will naturally want to breed her again if circumstances make breeding desirable and affordable.

On the other hand, if any part of your experience has been disappointing, you may prefer to sell the mare or possibly keep her for riding without breeding her again.

THE COLT AND HIS FUTURE

The colt is now ready to graduate to the more serious aspects of training, which will include flexion, use of the dumbjock, training in light harness, and later driving to a breaking cart, bagging, backing and possibly, if he shows a potential as a jumper, putting him over the low jumps of the cavaletti.

One point cannot be too strongly emphasized at this juncture and that is this: no training should be undertaken except by an expert.

We do not mean to imply by this statement that because you are very young, possibly no more than fifteen or sixteen years of age, that you cannot attempt the train-

MORE ADVANCED TRAINING

Who says a donkey foal isn't adorable?

ing of your own colt. We have known many young people, some naturally gifted, others trained by older horsemen, who have broken colts successfully. In fact frequently the lightness of the youthful rider, his patience, abundant love, and perseverance, may produce a far better result than an older less dedicated horseman might achieve.

If you have grown up around horses, understand that each one is different in disposition, character, and response to training, and if you are sufficiently humble to learn from your pupil as well as to instruct him, you may make an excellent trainer.

But for an inexperienced horseman to attempt to train an inexperienced horse would be for the blind to lead the blind, and only disaster could result.

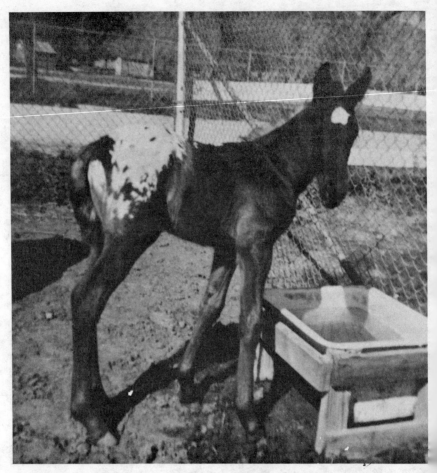

Safe watering set up for a young foal. (Courtesy Sylvia Martinez, R. W. Appaloosa Ranch, Saugus, California.)

Therefore if you want very much to train your own colt and you don't feel quite up to the project, we suggest that you call on an expert to break the colt for you and to teach you the fundamentals of training at the same time.

MORE ADVANCED TRAINING

THE ROLE OF THE TRAINER'S ASSISTANT

You may watch the training process and learn much from it, and you may eventually assist the instructor when you are ready to make a worthwhile contribution. But you should only contribute when asked to and at other times remain a silent observer. Failure to do this may result in injury to you or to the colt and may so confuse the animal that he will be unable to respond properly to correct commands.

The photographs, together with their accompanying text, will show steps in training mentioned above. They are offered as a means of illustrating the various steps to be taken in bringing the colt to the stage known as "green broke."

WHERE TO SEEK TRAINING HELP

If you wish to seek help in training your colt we suggest you contact your breed association if you own a registered mare. You may also contact your local 4-H Club. Or you may get in touch with your local riding clubs or equestrian groups.

10

AT THE SHOW

THE SUBTLE REWARDS OF FOSTER PARENTHOOD

Enjoying the beauty, the grace, the agility, and the pixie-like playfulness of your new colt is thrilling beyond the power of description. No one could count the number of pleasurable hours we have spent watching our young colts frisk and frolic at pasture, heads high, tails pointed skyward, prancing and dancing with all the bounce and spring of a deer.

You too will spend innumerable hours observing your colt, admiring its color, its style, its flair. This is your reward for all the time, money and effort you have expended. And what ample repayment it is.

But there is an even richer reward that can be yours—the moment when you take your darling to a show and hear the praise and approval bestowed upon him by strangers.

EXPOSED TO THE PUBLIC GAZE

The experience is like that of a young mother who has displayed her baby, with all its charms, to relatives and friends, basking the while in their vociferous acclaim.

Then the baby is finally curried and groomed—oops, pardon us—we meant to say bathed and adorned, and,

AT THE SHOW

propped in his pram, taken to the park.

There in competition with equally adorable well-groomed (rather well cared for) babies the pampered moppet is now exposed to the cold, appraising glances of supercilious nursemaids and envious mothers, each convinced that their offspring is more magnificent, handsome and talented than any other.

CONCEALED BLEMISHES REVEALED

How every blemish shows! A freckle, a cowlick, even a lopsided grin, though so engaging in the tolerant privacy of the family circle, now becomes the subject of disdainful and disparaging comment.

And thus it is when you take your colt to his first show.

Up to now you have been positive that yours was the most superb creature ever to appear on this earth. Friends and relatives (if not other horse owners) have pretended to agree with you—at least to your face.

But in the show ring, under the cold scrutiny of other horse owners and before the sharp appraising eyes of an unfeeling judge, how you quake and suffer.

The dish face you thought so adorable, the little habit of stomping a hoof in temper, suddenly appear far from enchanting.

Surely it isn't possible that the world could be filled with so many striking colts, all well-groomed, handsomely caparisoned and well behaved—except yours!

PRIDE TURNS TO CONSTERNATION, HOPE TO DESPAIR

It's a pattern with which every colt owner is familiar as pride turns to consternation, high hopes to despair. But it's a pattern that changes rapidly as both you and your colt gradually become more self-confident, less self-conscious in the show ring.

Young foal poses proudly for his picture. (Courtesy Pat Gast, Dancing Bear Ranch, Canyon Country, California.)

AT THE SHOW

To gain this savoir faire you begin your preparations for the show ring at home when the foal is about six months old.

FIRST STEPS TOWARD EXHIBITING

The first step is to accustom the foal to trailering for it is unlikely that he can reach any creditable show without being trailered there.

Force is unnecessary, in fact tabu, in teaching a young foal to accept the forbidding restraints of a horse trailer.

Here is a subtle way of luring him into a trailer that we have found works very well for us and it may work for you.

TRICKS TO MAKE TRAILERING ACCEPTABLE

Start with a capacious trailer anchored in a cool, quiet spot. Be sure the trailer is secured in such a way that it will not sway or creak when the foal enters it. The tailgate should be flat and level. There should be ample width and height to accommodate both the foal and its mother for this is the secret of success in coaxing the foal to enter this mysterious enclosure for the first time.

There is another secret and we offer it freely: place some sweet feed—enough for both mare and foal—in the feed bin of the trailer, making certain that the feed intended for the foal is at a height which he can reach comfortably and separated from the mare's feed so each can eat without scrabbling.

If the mare is accustomed to trailering, she should enter the trailer without hesitation, lured by the smell of the feed. You can then tie her and before long the foal, after much hesitation, attempted investigations, rapid retreats and false starts, will eventually (and this may take several days) join his mother in the trailer, and learn to eat happily beside her.

When this process has been repeated often enough to remove all fear, you may tie the colt after he becomes absorbed in his feed and remove his mother from the trailer. Of course this step should not be taken till he is thoroughly accustomed to the trailer and will not pull back, kick, or throw himself down at discovering that he is imprisoned in it.

When he is tied he must be snubbed up short and cross tied if possible to make sure that he won't become "cast" (so tightly jammed that he is unable to rise once he gets down).

SOME SHOW GENERALITIES

And now about show rules in general. For particulars we refer you to your local riding groups, breed associations, 4-H Clubs, and Horse Show Associations.

As we mentioned before, you can't enter your mare in any creditable horse show (with rare exceptions) unless she is registered. But there are frequently children's shows where grade mares may compete, and if such shows exist in your area, they will make an excellent starting point for a beginning exhibitor.

Since you will doubtless do your first showing of your colt with his dam, he should also be registered, and he can be registered with most breed associations as soon as he is sufficiently matured to establish his breed characteristics.

In the case of most breeds both parents must be registered if you hope to register their offspring. In the case of color lines such as the Palomino, Appaloosa, Pinto, Albino, etc., the sire must be registered, as explained previously, but the mare may very possibly be an unregistered grade mare.

AT THE SHOW

SHOWING THE MARE WITH FOAL AT SIDE

The first class you might consider entering is one usually designated as "Brood Mares—Foal at Side." Your local equestrian club or show association will supply entry blanks with rules for entering a show of this kind.

You will have to pay an entry fee and submit an entry blank properly filled out well in advance of show date, and you must supply all necessary data requested by the show committee.

Quarter horse mare and her three-month-old foal. (Courtesy Pat Gast, Dancing Bear Ranch, Canyon Country, California.)

THE EXCITEMENT OF PRE-SHOW PREPARATIONS

What activity there will be during those final days before show date as you put both mare and foal on special show or conditioning feed to round them out and sleek them up.

Special grooming begins well in advance of show date with particular attention given to washing, currying, brushing and polishing.

Manes and tails must be pulled (thinned) and plaited while damp to assure eye catching waves.

Hoofs must be trimmed and (at the last minute) oiled, to give them a lustrous sheen. In the case of the mare she should be neatly shod.

Tack (show halters, lead shanks, head stalls, etc.,) must be saddle soaped and polished till you can see your reflection in the glistening leather. Accents of brass must sparkle and gleam from endless applications of plain spit and elbow grease.

PERSONAL APPURTENANCES GIVE A PROFESSIONAL TOUCH

You'll want to assemble your show fittings in a little trunk, and by that we mean grooming aids and any other items you may want to take with you.

If you've the time, wherewithal, and energy you may want to design your own show blankets or cooling sheets featuring your own Stable Colors.

The trunk should be painted the same colors with your farm name and your horses' names stencilled on it. It should be large enough to hold blankets, tack, and grooming aids but not so bulky that you can't trailer it comfortably.

Stick-on felt appliqué letters may be applied in contrasting colors to blankets and cooling sheets spelling out your horse's name. Feed bags or buckets should also be handsomely decorated.

AT THE SHOW

The seasoned horseman takes kibbled feed to shows since it is easier to haul and handle than hay. But of course your mare and/or foal must have become accustomed to eating kibbled feed or you won't be successful at introducing it suddenly.

There's usually a pre-show party among horse lovers and a following celebration after the show if merited. This can be a happy custom for you to establish in your local horsey circle.

THE DAY OF THE SHOW

The day of the show you'll be up at dawn checking things out. This is the time for last minute grooming and the final assembling of all the items you've decided you'll need at the show.

Your check list should include feed, grooming aids, tack, blankets or cooling sheets, buckets, lead shanks, and tie ropes (bring plenty of these because they have a way of disappearing).

Then you want to be sure that your own show costume is trim and sparkling with cleanliness: boots shining, hat spotless, trousers or britches creased sharply. You'll hang your show garb in car or trailer and wear grub clothes in which to do your last minute grooming after you arrive at the show grounds.

Then when everything's squared away you'll switch to your show attire just before the entry parade.

ARRIVING AT THE SHOW GROUNDS

How exciting it is to arrive at the show grounds. The milling crowds, the hawkers, the music, the color, the smell of popcorn and hot dogs but above all the pungent, overpowering smell of horses, their shrill whinnies and nickerings and the knot at the pit of your stomach, all combine in one heady elixir of life that lifts you into a new realm of agony and anticipation, that makes you feel for the first time the true thrill of horse ownership.

We trust you have accustomed both mare and foal to the noise, the bustle, the movement, and the music in small doses well in advance of show date. If not, you will surely be in a sorry predicament if your mare throws a tizzy at the strangeness of her surroundings or your foal bolts for the blue horizon.

Nothing could be more embarrassing!

But if you have taken both mare and foal to unusual surroundings on several occasions prior to their debut and if you have played music on your own stereo to accustom them to the sound of it and if you make certain the giddy foal is allowed to stay close to his mother when first he hears the din and the clatter of the awesome show ring things may not turn out too badly.

We phrase that last sentence slightly in the negative because we don't want you to expect too much from a green colt, then you won't be disappointed. A first showing is really toe-dabbling in a new world for both of you. Do the best you can, be grateful for any honors achieved or just for the lessons learned if no honors happen to be forthcoming. Then start preparations for your next show.

ABOUT SPORTSMANSHIP

If ribbons are won, well and good. If none are won, don't let this disturb you. The ribbon or the award is not the important thing in showing. What really counts is the fun you have and the experience gained for you and your colt.

Don't do as too many young people do and become so competitive that the pleasure of showing your horse is spoiled for you unless you win awards.

We have seen friendships broken and families disrupted

and all the fun of horse ownership tarnished by too strong a feeling of competitiveness among young horse owners.

Far too often the adults or parents involved are to blame for this unhappy situation since they feel their own pride is injured if their children's entries fail to win highest honors. This is so deplorable we can't caution against it too strongly.

DON'T MISJUDGE THE JUDGE

And for heaven's sake don't begin to insist as so many horse owners do that judges are unfair or dishonest or unjust simply because your entry doesn't win a top prize.

Judges in the vast majority of instances—and we discount the exceptions—are dedicated horse lovers who devote many, many spare time hours to show judging simply to encourage good sportsmanship among young people and to keep the love of horses alive.

So give them the benefit of the doubt. If you think you should have won an award and someone else shouldn't have won it, shrug it off. Try again. Don't let your disposition or your pleasure be ruined. Next time you may be the winner and your competitor will have to accept the judge's decision with good grace.

If you keep plugging away, improving the performance of your colt as you improve your own mastery of horses, and if you keep entering shows with good humor and persistence you will find that nothing can keep you from winning the awards you deserve.

And while you are at the show, mind your manners. Observe the rules of courtesy entering and leaving the ring and during exhibition. Most rules are made for your safety or the safety of your horse, so don't gripe about them. Instead find out the reason behind the rule and you'll decide that you want to obey it.

On leaving the show for home check over everything once again to be sure you're taking home whatever you started out with.

SHOWING THE COLT ON HIS OWN

After you have shown your mare with foal at side there will come a time, usually when the colt is a year old or older, when you will want to show him on his own in an "In Halter" Class.

You will follow the same procedure for entering him as previously mentioned but you will prepare for this exhibition by teaching him to carry out the following duties.

1) With lead shank attached to his halter, teach him to walk sedately back and forth in front of an imaginary judge's stand.

2) Teach him to follow you at the trot as you run ahead of him, leading him at this faster pace, back and forth as if for the judge's inspection. His head and tail should be up, his leg action high and pronounced, his gait rapid but *controlled*.

3) Teach him to stand "stretched," head erect, feet squarely planted and motionless as the judge checks him for conformation.

After showing in halter your colt must be "finished" before he can be exhibited under saddle.

This means he must learn to respond to a training bit (after preliminary riding with the bitless bridle or hackamore); he must guide, neck rein, back up, come to a sliding halt (in the case of a western stock horse), he must behave creditably at the walk, trot (jog), or canter (lope), while being ridden in a circle to the left and to the right.

He must respond to the aids by leading on the right or

left diagonal at the trot, the near or off side at the canter, depending on which direction he is being ridden in a circle.

And if he shows potential as a jumper, he should be able to clear a three foot jump in a level-headed and controlled manner.

A LOOK TO THE FUTURE

Having brought your first colt to the show stage, we have accomplished all we set out to accomplish within the pages of this particular book.

Now that you've come with us all the way we feel certain that you will agree that you have learned a great deal about horses as the result of this adventure in parenthood that you couldn't have learned any other way.

You have suffered the tedium of gestation and the pangs of birth; you have enjoyed the wonder of motherhood and the joys of babyhood; you have shared the disciplines of schooling and sampled the thrills of the show ring—what greater thrill could lie ahead?

To that question we always give the same answer: there's only one thing in this world that we guarantee can be more fun than raising a colt—and you guessed it— that's raising a second colt.

So why not get started right now?

A PERSONAL WORD FROM MELVIN POWERS
PUBLISHER, WILSHIRE BOOK COMPANY

Dear Friend:

My goal is to publish interesting, informative, and inspirational books. You can help me accomplish this by answering the following questions, either by phone or by mail. Or, if convenient for you, I would welcome the opportunity to visit with you in my office and hear your comments in person.

Did you enjoy reading this book? Why?

Would you enjoy reading another similar book?

What idea in the book impressed you the most?

If applicable to your situation, have you incorporated this idea in your daily life?

Is there a chapter that could serve as a theme for an entire book? Please explain.

If you have an idea for a book, I would welcome discussing it with you. If you already have one in progress, write or call me concerning possible publication. I can be reached at (213) 875-1711 or (213) 983-1105.

Sincerely yours,

MELVIN POWERS

12015 Sherman Road
North Hollywood, California 91605

MELVIN POWERS SELF-IMPROVEMENT LIBRARY

ASTROLOGY

ASTROLOGY: A FASCINATING HISTORY *P. Naylor*	2.00
ASTROLOGY: HOW TO CHART YOUR HOROSCOPE *Max Heindel*	3.00
ASTROLOGY: YOUR PERSONAL SUN-SIGN GUIDE *Beatrice Ryder*	3.00
ASTROLOGY FOR EVERYDAY LIVING *Janet Harris*	2.00
ASTROLOGY MADE EASY *Astarte*	2.00
ASTROLOGY MADE PRACTICAL *Alexandra Kayhle*	3.00
ASTROLOGY, ROMANCE, YOU AND THE STARS *Anthony Norvell*	4.00
MY WORLD OF ASTROLOGY *Sydney Omarr*	4.00
THOUGHT DIAL *Sydney Omarr*	3.00
ZODIAC REVEALED *Rupert Gleadow*	2.00

BRIDGE

BRIDGE BIDDING MADE EASY *Edwin B. Kantar*	5.00
BRIDGE CONVENTIONS *Edwin B. Kantar*	4.00
BRIDGE HUMOR *Edwin B. Kantar*	3.00
COMPETITIVE BIDDING IN MODERN BRIDGE *Edgar Kaplan*	4.00
DEFENSIVE BRIDGE PLAY COMPLETE *Edwin B. Kantar*	10.00
HOW TO IMPROVE YOUR BRIDGE *Alfred Sheinwold*	2.00
INTRODUCTION TO DEFENDER'S PLAY *Edwin B. Kantar*	3.00
TEST YOUR BRIDGE PLAY *Edwin B. Kantar*	3.00
WINNING DECLARER PLAY *Dorothy Hayden Truscott*	4.00

BUSINESS, STUDY & REFERENCE

CONVERSATION MADE EASY *Elliot Russell*	2.00
EXAM SECRET *Dennis B. Jackson*	2.00
FIX-IT BOOK *Arthur Symons*	2.00
HOW TO DEVELOP A BETTER SPEAKING VOICE *M. Hellier*	2.00
HOW TO MAKE A FORTUNE IN REAL ESTATE *Albert Winnikoff*	3.00
INCREASE YOUR LEARNING POWER *Geoffrey A. Dudley*	2.00
MAGIC OF NUMBERS *Robert Tocquet*	2.00
PRACTICAL GUIDE TO BETTER CONCENTRATION *Melvin Powers*	2.00
PRACTICAL GUIDE TO PUBLIC SPEAKING *Maurice Forley*	3.00
7 DAYS TO FASTER READING *William S. Schaill*	3.00
SONGWRITERS RHYMING DICTIONARY *Jane Shaw Whitfield*	5.00
SPELLING MADE EASY *Lester D. Basch & Dr. Milton Finkelstein*	2.00
STUDENT'S GUIDE TO BETTER GRADES *J. A. Rickard*	2.00
TEST YOURSELF—Find Your Hidden Talent *Jack Shafer*	2.00
YOUR WILL & WHAT TO DO ABOUT IT *Attorney Samuel G. Kling*	3.00

CALLIGRAPHY

CALLIGRAPHY—The Art of Beautfiul Writing *Katherine Jeffares*	5.00

CHESS & CHECKERS

BEGINNER'S GUIDE TO WINNING CHESS *Fred Reinfeld*	3.00
BETTER CHESS—How to Play *Fred Reinfeld*	2.00
CHECKERS MADE EASY *Tom Wiswell*	2.00
CHESS IN TEN EASY LESSONS *Larry Evans*	3.00
CHESS MADE EASY *Milton L. Hanauer*	2.00
CHESS MASTERY—A New Approach *Fred Reinfeld*	2.00
CHESS PROBLEMS FOR BEGINNERS *edited by Fred Reinfeld*	2.00
CHESS SECRETS REVEALED *Fred Reinfeld*	2.00
CHESS STRATEGY—An Expert's Guide *Fred Reinfeld*	2.00
CHESS TACTICS FOR BEGINNERS *edited by Fred Reinfeld*	2.00
CHESS THEORY & PRACTICE *Morry & Mitchell*	2.00
HOW TO WIN AT CHECKERS *Fred Reinfeld*	2.00
1001 BRILLIANT WAYS TO CHECKMATE *Fred Reinfeld*	3.00
1001 WINNING CHESS SACRIFICES & COMBINATIONS *Fred Reinfeld*	3.00
SOVIET CHESS *Edited by R. G. Wade*	3.00

COOKERY & HERBS

CULPEPER'S HERBAL REMEDIES *Dr. Nicholas Culpeper*	2.00
FAST GOURMET COOKBOOK *Poppy Cannon*	2.50
HEALING POWER OF HERBS *May Bethel*	3.00

____	HEALING POWER OF NATURAL FOODS *May Bethel*	3.00
____	HERB HANDBOOK *Dawn MacLeod*	2.00
____	HERBS FOR COOKING AND HEALING *Dr. Donald Law*	2.00
____	HERBS FOR HEALTH—How to Grow & Use Them *Louise Evans Doole*	2.00
____	HOME GARDEN COOKBOOK—Delicious Natural Food Recipes *Ken Kraft*	3.00
____	MEDICAL HERBALIST *edited by Dr. J. R. Yemm*	3.00
____	NATURAL FOOD COOKBOOK *Dr. Harry C. Bond*	3.00
____	NATURE'S MEDICINES *Richard Lucas*	3.00
____	VEGETABLE GARDENING FOR BEGINNERS *Hugh Wiberg*	2.00
____	VEGETABLES FOR TODAY'S GARDENS *R. Milton Carleton*	2.00
____	VEGETARIAN COOKERY *Janet Walker*	3.00
____	VEGETARIAN COOKING MADE EASY & DELECTABLE *Veronica Vezza*	2.00
____	VEGETARIAN DELIGHTS—A Happy Cookbook for Health *K. R. Mehta*	2.00
____	VEGETARIAN GOURMET COOKBOOK *Joyce McKinnel*	3.00

GAMBLING & POKER

____	ADVANCED POKER STRATEGY & WINNING PLAY *A. D. Livingston*	3.00
____	HOW NOT TO LOSE AT POKER *Jeffrey Lloyd Castle*	3.00
____	HOW TO WIN AT DICE GAMES *Skip Frey*	3.00
____	HOW TO WIN AT POKER *Terence Reese & Anthony T. Watkins*	2.00
____	SECRETS OF WINNING POKER *George S. Coffin*	3.00
____	WINNING AT CRAPS *Dr. Lloyd T. Commins*	2.00
____	WINNING AT GIN *Chester Wander & Cy Rice*	3.00
____	WINNING AT POKER—An Expert's Guide *John Archer*	3.00
____	WINNING AT 21—An Expert's Guide *John Archer*	3.00
____	WINNING POKER SYSTEMS *Norman Zadeh*	3.00

HEALTH

____	DR. LINDNER'S SPECIAL WEIGHT CONTROL METHOD	1.50
____	HELP YOURSELF TO BETTER SIGHT *Margaret Darst Corbett*	3.00
____	HOW TO IMPROVE YOUR VISION *Dr. Robert A. Kraskin*	2.00
____	HOW YOU CAN STOP SMOKING PERMANENTLY *Ernest Caldwell*	2.00
____	MIND OVER PLATTER *Peter G. Lindner, M.D.*	2.00
____	NATURE'S WAY TO NUTRITION & VIBRANT HEALTH *Robert J. Scrutton*	3.00
____	NEW CARBOHYDRATE DIET COUNTER *Patti Lopez-Pereira*	1.50
____	PSYCHEDELIC ECSTASY *William Marshall & Gilbert W. Taylor*	2.00
____	REFLEXOLOGY *Dr. Maybelle Segal*	2.00
____	YOU CAN LEARN TO RELAX *Dr. Samuel Gutwirth*	2.00
____	YOUR ALLERGY—What To Do About It *Allan Knight, M.D.*	2.00

HOBBIES

____	BATON TWIRLING—A Complete Illustrated Guide *Doris Wheelus*	4.00
____	BEACHCOMBING FOR BEGINNERS *Norman Hickin*	2.00
____	BLACKSTONE'S MODERN CARD TRICKS *Harry Blackstone*	2.00
____	BLACKSTONE'S SECRETS OF MAGIC *Harry Blackstone*	2.00
____	BUTTERFLIES	2.50
____	COIN COLLECTING FOR BEGINNERS *Burton Hobson & Fred Reinfeld*	2.00
____	ENTERTAINING WITH ESP *Tony 'Doc' Shiels*	2.00
____	400 FASCINATING MAGIC TRICKS YOU CAN DO *Howard Thurston*	3.00
____	GOULD'S GOLD & SILVER GUIDE TO COINS *Maurice Gould*	2.00
____	HOW I TURN JUNK INTO FUN AND PROFIT *Sari*	3.00
____	HOW TO PLAY THE HARMONICA FOR FUN AND PROFIT *Hal Leighton*	3.00
____	HOW TO WRITE A HIT SONG & SELL IT *Tommy Boyce*	7.00
____	JUGGLING MADE EASY *Rudolf Dittrich*	2.00
____	MAGIC MADE EASY *Byron Wels*	2.00
____	STAMP COLLECTING FOR BEGINNERS *Burton Hobson*	2.00
____	STAMP COLLECTING FOR FUN & PROFIT *Frank Cetin*	2.00

HORSE PLAYERS' WINNING GUIDES

____	BETTING HORSES TO WIN *Les Conklin*	3.00
____	ELIMINATE THE LOSERS *Bob McKnight*	3.00
____	HOW TO PICK WINNING HORSES *Bob McKnight*	3.00
____	**HOW TO WIN AT THE RACES** *Sam (The Genius) Lewin*	3.00
____	**HOW YOU CAN BEAT THE RACES** *Jack Kavanagh*	3.00
____	**MAKING MONEY AT THE RACES** *David Barr*	3.00

___	PAYDAY AT THE RACES *Les Conklin*	2.00
___	SMART HANDICAPPING MADE EASY *William Bauman*	3.00
___	SUCCESS AT THE HARNESS RACES *Barry Meadow*	2.50
___	WINNING AT THE HARNESS RACES—An Expert's Guide *Nick Cammarano*	3.00

HUMOR

___	HOW TO BE A COMEDIAN FOR FUN & PROFIT *King & Laufer*	2.00
___	JOKE TELLER'S HANDBOOK *Bob Orben*	3.00

HYPNOTISM

___	ADVANCED TECHNIQUES OF HYPNOSIS *Melvin Powers*	2.00
___	BRAINWASHING AND THE CULTS *Paul A. Verdier, Ph.D.*	3.00
___	CHILDBIRTH WITH HYPNOSIS *William S. Kroger, M.D.*	3.00
___	HOW TO SOLVE Your Sex Problems with Self-Hypnosis *Frank S. Caprio, M.D.*	3.00
___	HOW TO STOP SMOKING THRU SELF-HYPNOSIS *Leslie M. LeCron*	2.00
___	HOW TO USE AUTO-SUGGESTION EFFECTIVELY *John Duckworth*	3.00
___	HOW YOU CAN BOWL BETTER USING SELF-HYPNOSIS *Jack Heise*	2.00
___	HOW YOU CAN PLAY BETTER GOLF USING SELF-HYPNOSIS *Jack Heise*	2.00
___	HYPNOSIS AND SELF-HYPNOSIS *Bernard Hollander, M.D.*	3.00
___	HYPNOTISM (Originally published in 1893) *Carl Sextus*	3.00
___	HYPNOTISM & PSYCHIC PHENOMENA *Simeon Edmunds*	3.00
___	HYPNOTISM MADE EASY *Dr. Ralph Winn*	3.00
___	HYPNOTISM MADE PRACTICAL *Louis Orton*	2.00
___	HYPNOTISM REVEALED *Melvin Powers*	2.00
___	HYPNOTISM TODAY *Leslie LeCron and Jean Bordeaux, Ph.D.*	4.00
___	MODERN HYPNOSIS *Lesley Kuhn & Salvatore Russo, Ph.D.*	4.00
___	NEW CONCEPTS OF HYPNOSIS *Bernard C. Gindes, M.D.*	4.00
___	NEW SELF-HYPNOSIS *Paul Adams*	3.00
___	POST-HYPNOTIC INSTRUCTIONS—Suggestions for Therapy *Arnold Furst*	3.00
___	PRACTICAL GUIDE TO SELF-HYPNOSIS *Melvin Powers*	2.00
___	PRACTICAL HYPNOTISM *Philip Magonet, M.D.*	2.00
___	SECRETS OF HYPNOTISM *S. J. Van Pelt, M.D.*	3.00
___	SELF-HYPNOSIS Its Theory, Technique & Application *Melvin Powers*	2.00
___	SELF-HYPNOSIS A Conditioned-Response Technique *Laurance Sparks*	4.00
___	THERAPY THROUGH HYPNOSIS edited by *Raphael H. Rhodes*	3.00

JUDAICA

___	HOW TO LIVE A RICHER & FULLER LIFE *Rabbi Edgar F. Magnin*	2.00
___	MODERN ISRAEL *Lily Edelman*	2.00
___	OUR JEWISH HERITAGE *Rabbi Alfred Wolf & Joseph Gaer*	2.00
___	ROMANCE OF HASSIDISM *Jacob S. Minkin*	2.50
___	SERVICE OF THE HEART *Evelyn Garfiel, Ph.D.*	4.00
___	STORY OF ISRAEL IN COINS *Jean & Maurice Gould*	2.00
___	STORY OF ISRAEL IN STAMPS *Maxim & Gabriel Shamir*	1.00
___	TONGUE OF THE PROPHETS *Robert St. John*	3.00
___	TREASURY OF COMFORT edited by *Rabbi Sidney Greenberg*	4.00

JUST FOR WOMEN

___	COSMOPOLITAN'S GUIDE TO MARVELOUS MEN Fwd. by *Helen Gurley Brown*	3.00
___	COSMOPOLITAN'S NEW ETIQUETTE GUIDE Fwd. by *Helen Gurley Brown*	4.00
___	COSMOPOLITAN'S HANG-UP HANDBOOK Foreword by *Helen Gurley Brown*	4.00
___	COSMOPOLITAN'S LOVE BOOK—A Guide to Ecstasy in Bed	3.00
___	JUST FOR WOMEN—A Guide to the Female Body *Richard E. Sand, M.D.*	3.00
___	NEW APPROACHES TO SEX IN MARRIAGE *John E. Eichenlaub, M.D.*	3.00
___	SEXUALLY ADEQUATE FEMALE *Frank S. Caprio, M.D.*	2.00
___	YOUR FIRST YEAR OF MARRIAGE *Dr. Tom McGinnis*	3.00

MARRIAGE, SEX & PARENTHOOD

___	ABILITY TO LOVE *Dr. Allan Fromme*	5.00
___	ENCYCLOPEDIA OF MODERN SEX & LOVE TECHNIQUES *Macandrew*	4.00
___	GUIDE TO SUCCESSFUL MARRIAGE *Drs. Albert Ellis & Robert Harper*	3.00
___	HOW TO RAISE AN EMOTIONALLY HEALTHY, HAPPY CHILD *A. Ellis*	3.00
___	IMPOTENCE & FRIGIDITY *Edwin W. Hirsch, M.D.*	3.00
___	SEX WITHOUT GUILT *Albert Ellis, Ph.D.*	3.00
___	SEXUALLY ADEQUATE MALE *Frank S. Caprio, M.D.*	3.00

METAPHYSICS & OCCULT

	BOOK OF TALISMANS, AMULETS & ZODIACAL GEMS *William Pavitt*	4.00
	CONCENTRATION—A Guide to Mental Mastery *Mouni Sadhu*	3.00
	CRITIQUES OF GOD *Edited by Peter Angeles*	7.00
	DREAMS & OMENS REVEALED *Fred Gettings*	2.00
	EXTRASENSORY PERCEPTION *Simeon Edmunds*	2.00
	EXTRA-TERRESTRIAL INTELLIGENCE—The First Encounter	6.00
	FORTUNE TELLING WITH CARDS *P. Foli*	2.00
	HANDWRITING ANALYSIS MADE EASY *John Marley*	2.00
	HANDWRITING TELLS *Nadya Olyanova*	5.00
	HOW TO UNDERSTAND YOUR DREAMS *Geoffrey A. Dudley*	2.00
	ILLUSTRATED YOGA *William Zorn*	3.00
	IN DAYS OF GREAT PEACE *Mouni Sadhu*	3.00
	KING SOLOMON'S TEMPLE IN THE MASONIC TRADITION *Alex Horne*	5.00
	LSD—THE AGE OF MIND *Bernard Roseman*	2.00
	MAGICIAN—His training and work *W. E. Butler*	2.00
	MEDITATION *Mouni Sadhu*	4.00
	MODERN NUMEROLOGY *Morris C. Goodman*	3.00
	NUMEROLOGY—ITS FACTS AND SECRETS *Ariel Yvon Taylor*	2.00
	PALMISTRY MADE EASY *Fred Gettings*	3.00
	PALMISTRY MADE PRACTICAL *Elizabeth Daniels Squire*	3.00
	PALMISTRY SECRETS REVEALED *Henry Frith*	2.00
	PRACTICAL YOGA *Ernest Wood*	3.00
	PROPHECY IN OUR TIME *Martin Ebon*	2.50
	PSYCHOLOGY OF HANDWRITING *Nadya Olyanova*	3.00
	SEEING INTO THE FUTURE *Harvey Day*	2.00
	SUPERSTITION—Are you superstitious? *Eric Maple*	2.00
	TAROT *Mouni Sadhu*	5.00
	TAROT OF THE BOHEMIANS *Papus*	5.00
	TEST YOUR ESP *Martin Ebon*	2.00
	WAYS TO SELF-REALIZATION *Mouni Sadhu*	3.00
	WITCHCRAFT, MAGIC & OCCULTISM—A Fascinating History *W. B. Crow*	3.00
	WITCHCRAFT—THE SIXTH SENSE *Justine Glass*	2.00
	WORLD OF PSYCHIC RESEARCH *Hereward Carrington*	2.00
	YOU CAN ANALYZE HANDWRITING *Robert Holder*	2.00

SELF-HELP & INSPIRATIONAL

	CYBERNETICS WITHIN US *Y. Saparina*	3.00
	DAILY POWER FOR JOYFUL LIVING *Dr. Donald Curtis*	2.00
	DOCTOR PSYCHO-CYBERNETICS *Maxwell Maltz, M.D.*	3.00
	DYNAMIC THINKING *Melvin Powers*	2.00
	EXUBERANCE—Your Guide to Happiness & Fulfillment *Dr. Paul Kurtz*	3.00
	GREATEST POWER IN THE UNIVERSE *U. S. Andersen*	4.00
	GROW RICH WHILE YOU SLEEP *Ben Sweetland*	3.00
	GROWTH THROUGH REASON *Albert Ellis, Ph.D.*	4.00
	GUIDE TO DEVELOPING YOUR POTENTIAL *Herbert A. Otto, Ph.D.*	3.00
	GUIDE TO LIVING IN BALANCE *Frank S. Caprio, M.D.*	2.00
	HELPING YOURSELF WITH APPLIED PSYCHOLOGY *R. Henderson*	2.00
	HELPING YOURSELF WITH PSYCHIATRY *Frank S. Caprio, M.D.*	2.00
	HOW TO ATTRACT GOOD LUCK *A. H. Z. Carr*	3.00
	HOW TO CONTROL YOUR DESTINY *Norvell*	3.00
	HOW TO DEVELOP A WINNING PERSONALITY *Martin Panzer*	3.00
	HOW TO DEVELOP AN EXCEPTIONAL MEMORY *Young & Gibson*	4.00
	HOW TO OVERCOME YOUR FEARS *M. P. Leahy, M.D.*	2.00
	HOW YOU CAN HAVE CONFIDENCE AND POWER *Les Giblin*	3.00
	HUMAN PROBLEMS & HOW TO SOLVE THEM *Dr. Donald Curtis*	3.00
	I CAN *Ben Sweetland*	4.00
	I WILL *Ben Sweetland*	3.00
	LEFT-HANDED PEOPLE *Michael Barsley*	3.00
	MAGIC IN YOUR MIND *U. S. Andersen*	4.00
	MAGIC OF THINKING BIG *Dr. David J. Schwartz*	3.00
	MAGIC POWER OF YOUR MIND *Walter M. Germain*	4.00

___	MENTAL POWER THROUGH SLEEP SUGGESTION *Melvin Powers*	2.00
___	NEW GUIDE TO RATIONAL LIVING *Albert Ellis, Ph.D. & R. Harper, Ph.D.*	3.00
___	OUR TROUBLED SELVES *Dr. Allan Fromme*	3.00
___	PRACTICAL GUIDE TO SUCCESS & POPULARITY *C. W. Bailey*	2.00
___	PSYCHO-CYBERNETICS *Maxwell Maltz, M.D.*	2.00
___	SCIENCE OF MIND IN DAILY LIVING *Dr. Donald Curtis*	2.00
___	SECRET POWER OF THE PYRAMIDS *U. S. Andersen*	4.00
___	SECRET OF SECRETS *U. S. Andersen*	4.00
___	STUTTERING AND WHAT YOU CAN DO ABOUT IT *W. Johnson, Ph.D.*	2.50
___	SUCCESS-CYBERNETICS *U. S. Andersen*	4.00
___	10 DAYS TO A GREAT NEW LIFE *William E. Edwards*	3.00
___	THINK AND GROW RICH *Napoleon Hill*	3.00
___	THREE MAGIC WORDS *U. S. Andersen*	4.00
___	TREASURY OF THE ART OF LIVING *Sidney S. Greenberg*	5.00
___	YOU ARE NOT THE TARGET *Laura Huxley*	3.00
___	YOUR SUBCONSCIOUS POWER *Charles M. Simmons*	4.00
___	YOUR THOUGHTS CAN CHANGE YOUR LIFE *Dr. Donald Curtis*	3.00

SPORTS

___	ARCHERY—An Expert's Guide *Dan Stamp*	2.00
___	BICYCLING FOR FUN AND GOOD HEALTH *Kenneth E. Luther*	2.00
___	BILLIARDS—Pocket • Carom • Three Cushion *Clive Cottingham, Jr.*	2.00
___	CAMPING-OUT 101 Ideas & Activities *Bruno Knobel*	2.00
___	COMPLETE GUIDE TO FISHING *Vlad Evanoff*	2.00
___	HOW TO WIN AT POCKET BILLIARDS *Edward D. Knuchell*	3.00
___	LEARNING & TEACHING SOCCER SKILLS *Eric Worthington*	3.00
___	MOTORCYCLING FOR BEGINNERS *I. G. Edmonds*	2.00
___	PRACTICAL BOATING *W. S. Kals*	3.00
___	RACQUETBALL MADE EASY *Steve Lubarsky, Rod Delson & Jack Scagnetti*	3.00
___	SECRET OF BOWLING STRIKES *Dawson Taylor*	2.00
___	SECRET OF PERFECT PUTTING *Horton Smith & Dawson Taylor*	3.00
___	SECRET WHY FISH BITE *James Westman*	2.00
___	SKIER'S POCKET BOOK *Otti Wiedman* (4¼" x 6")	2.50
___	SOCCER—The game & how to play it *Gary Rosenthal*	2.00
___	STARTING SOCCER *Edward F. Dolan, Jr.*	2.00
___	TABLE TENNIS MADE EASY *Johnny Leach*	2.00

TENNIS LOVERS' LIBRARY

___	BEGINNER'S GUIDE TO WINNING TENNIS *Helen Hull Jacobs*	2.00
___	HOW TO BEAT BETTER TENNIS PLAYERS *Loring Fiske*	4.00
___	HOW TO IMPROVE YOUR TENNIS—Style, Strategy & Analysis *C. Wilson*	2.00
___	INSIDE TENNIS—Techniques of Winning *Jim Leighton*	3.00
___	PLAY TENNIS WITH ROSEWALL *Ken Rosewall*	2.00
___	PSYCH YOURSELF TO BETTER TENNIS *Dr. Walter A. Luszki*	2.00
___	SUCCESSFUL TENNIS *Neale Fraser*	2.00
___	TENNIS FOR BEGINNERS *Dr. H. A. Murray*	2.00
___	TENNIS MADE EASY *Joel Brecheen*	2.00
___	WEEKEND TENNIS—How to have fun & win at the same time *Bill Talbert*	3.00
___	WINNING WITH PERCENTAGE TENNIS—Smart Strategy *Jack Lowe*	2.00

WILSHIRE PET LIBRARY

___	DOG OBEDIENCE TRAINING *Gust Kessopulos*	3.00
___	DOG TRAINING MADE EASY & FUN *John W. Kellogg*	2.00
___	HOW TO BRING UP YOUR PET DOG *Kurt Unkelbach*	2.00
___	HOW TO RAISE & TRAIN YOUR PUPPY *Jeff Griffen*	2.00
___	PIGEONS: HOW TO RAISE & TRAIN THEM *William H. Allen, Jr.*	2.00

The books listed above can be obtained from your book dealer or directly from Melvin Powers. When ordering, please remit 30¢ per book postage & handling. Send for our free illustrated catalog of self-improvement books.

Melvin Powers
12015 Sherman Road, No. Hollywood, California 91605

Melvin Powers' Favorite Books

EXUBERANCE
Your Personal Guide to Happiness & Fulfillment
by Paul Kurtz, Ph.D.

Here is an outstanding book that can add more joy, meaning and zest to your life. Dr. Paul Kurtz, a professor of philosophy, gives you a guide to achieving your potential for happiness, love and fulfillment — even beyond your greatest expectations. He details how to add exuberance to each day and how to experience the good life permanently.
176 Pages ... $3

HOW YOU CAN HAVE CONFIDENCE & POWER IN DEALING WITH PEOPLE
by Les Giblin

Contents: 1. Your Key to Success and Happiness 2. How to Use the Basic Secret for Influencing Others 3. How to Cash in on Your Hidden Assets 4. How to Control the Actions & Attitudes of Others 5. How You Can Create a Good Impression on Other People 6. Techniques for Making & Keeping Friends 7. How to Use Three Big Secrets for Attracting People 8. How to Make the Other Person Feel Friendly — Instantly 9. How You Can Develop Skill in Using Words 10. The Technique of "White Magic" 11. How to Get Others to See Things Your Way — Quickly 12. A Simple, Effective Plan of Action That Will Bring You Success and Happiness.
192 Pages ... $3

A NEW GUIDE TO RATIONAL LIVING
by Albert Ellis, Ph.D. & Robert A. Harper, Ph.D.

Contents: 1. How Far Can You Go With Self-Analysis? 2. You Feel the Way You Think 3. Feeling Well by Thinking Straight 4. How You Create Your Feelings 5. Thinking Yourself Out of Emotional Disturbances 6. Recognizing and Attacking Neurotic Behavior 7. Overcoming the Influences of the Past 8. Does Reason Always Prove Reasonable? 9. Refusing to Feel Desperately Unhappy 10. Tackling Dire Needs for Approval 11. Eradicating Dire Fears of Failure 12. How to Stop Blaming and Start Living 13. How to Feel Undepressed though Frustrated 14. Controlling Your Own Destiny 15. Conquering Anxiety 16. Acquiring Self-discipline 17. Rewriting Your Personal History 18. Accepting Reality 19. Overcoming Inertia and Getting Creatively Absorbed 20. Living Rationally in an Irrational World 21. Rational-Emotive Therapy or Rational Behavior Training Updated.
256 Pages ... $3

PSYCHO-CYBERNETICS
A New Technique for Using Your Subconscious Power
by Maxwell Maltz, M.D., F.I.C.S.

Contents: 1. The Self Image: Your Key to a Better Life 2. Discovering the Success Mechanism within You 3. Imagination — The First Key to Your Success Mechanism 4. Dehypnotize Yourself from False Beliefs 5. How to Utilize the Power of Rational Thinking 6. Relax and Let Your Success Mechanism Work for You 7. You Can Acquire the Habit of Happiness 8. Ingredients of the Success-Type Personality and How to Acquire Them 9. The Failure Mechanism: How to Make It Work For You Instead of Against You 10. How to Remove Emotional Scars, or How to Give Yourself an Emotional Face Lift 11. How to Unlock Your Real Personality 12. Do-It-Yourself Tranquilizers That Bring Peace of Mind 13. How to Turn a Crisis into a Creative Opportunity 14. How to Get "That Winning Feeling" 15. More Years of Life — More Life in Your Years.
288 Pages ... $2

A PRACTICAL GUIDE TO SELF-HYPNOSIS
by Melvin Powers

Contents: 1. What You Should Know about Self-Hypnosis 2. What about the Dangers of Hypnosis? 3. Is Hypnosis the Answer? 4. How Does Self-Hypnosis Work? 5. How to Arouse Yourself from the Self-Hypnotic State 6. How to Attain Self-Hypnosis 7. Deepening the Self-Hypnotic State 8. What You Should Know about Becoming an Excellent Subject 9. Techniques for Reaching the Somnambulistic State 10. A New Approach to Self-Hypnosis When All Else Fails 11. Psychological Aids and Their Function 12. The Nature of Hypnosis 13. Practical Applications of Self-Hypnosis.
128 Pages ... $2

The books listed above can be obtained from your book dealer or directly from Melvin Powers. When ordering, please remit 30¢ per book postage & handling. Send for our free illustrated catalog of self-improvement books.

Melvin Powers
12015 Sherman Road, No. Hollywood, California 91605